MW00573499

IMPLEMENTING POSITIVE ORGANIZATIONAL CHANGE

A STRATEGIC PROJECT MANAGEMENT APPROACH

GINA ABUDI, MBA

Copyright © 2017 by Gina Abudi

ISBN-13: 978-1-60427-133-1

Printed and bound in the U.S.A. Printed on acid-free paper.

10 9 8 7 6 5 4 3 2 1

Library of Congress Cataloging-in-Publication Data

Names: Abudi, Gina, author.
   Title: Implementing positive organizational change : a strategic project
management approach / by Gina Abudi.
   Description: Plantation, FL: J. Ross Publishing, [2017] | Includes index.
   Identifiers: LCCN 2016052910 (print) | LCCN 2017002780 (ebook) |
ISBN 9781604271331 (hardcover : alk. paper) | ISBN 9781604277777
(e-book)
   Subjects: LCSH: Organizational change—Management. | Project
management.
   Classification: LCC HD58.8 .A284 2017 (print) | LCC HD58.8
(ebook) | DDC
   658.4/06—dc23

This publication contains information obtained from authentic and highly re-
garded sources. Reprinted material is used with permission, and sources are
indicated. Reasonable effort has been made to publish reliable data and information, but the author and the publisher cannot assume responsibility for the
validity of all materials or for the consequences of their use.

All rights reserved. Neither this publication nor any part thereof may be
reproduced, stored in a retrieval system, or transmitted in any form or by any
means, electronic, mechanical, photocopying, recording or otherwise, without
the prior written permission of the publisher.

The copyright owner's consent does not extend to copying for general distribution for promotion, for creating new works, or for resale. Specific permission must be obtained from J. Ross Publishing for such purposes.

Direct all inquiries to J. Ross Publishing, Inc., 300 S. Pine Island Rd., Suite
305, Plantation, FL 33324.

Phone: (954) 727-9333
Fax: (561) 892-0700
Web: www.jrosspub.com

CONTENTS

PREFACE

Change is a fact of life and is happening all around us every day—both personally and in our professional lives. But change is becoming more challenging, especially within organizations. Change is frequently launched when an organization has their proverbial back up against the wall, and they have no choice but to change. This might occur because the organization feels pain due to:

- Customers moving over to the competition
- New competition heating up the marketplace
- Products and services that are no longer meeting customer needs
- New regulations that must be implemented
- Top talent walking out the door
- Decreased profitability and reduced revenues
- Takeover threats

When an organization changes *only* when it is forced to do so, change is viewed as a negative—something that should be avoided unless it is necessary. In fact, change should be seen as a positive—an opportunity to create something new and exciting and valuable. When leaders engage people in conversations around change *before* we need to change, change is then seen as a positive—something to be embraced—rather than a negative that must be avoided. This doesn't mean that every individual will embrace change. As human beings, change is worrisome for us; it changes the status quo. However, by engaging people in change early on and throughout the change initiative, leadership is likely to have more champions and fewer resisters.

This book is needed in the marketplace because change is *not* going away and because it is a fact of business. It is becoming more essential for organizations to continuously change in today's increasingly competitive global marketplace. Those organizations that do not regularly undertake change initiatives will find it difficult, if not impossible, to continue to be successful. With all the change that happens, one would think it would be successful. That is not the case! Studies show that there is a 60-70% failure rate for organizational change projects. This statistic has stayed consistent since the 1970s. While there are significant concepts, theories, and methodologies around change, leadership capability to launch and implement successful change is often ineffective and contributes to that failure rate.

Change initiatives that are launched in the organization are often focused on the benefit of the change *to the organization*. The people-side of change is often forgotten. Without people, however, change initiatives *cannot* be successful.

This book will share a number of best practices for engaging people in change initiatives from early on and throughout the implementation of the change initiative, so as to ensure increased adoption of the change and for the change to *stick* over the long term. Through understanding the barriers to change as well as the impact of change on the organization and the individual employee, readers will be far better prepared to launch and work on organizational change initiatives while engaging employees throughout the organization. Through the application of best practices, tips, and resources provided in the book, readers will not only increase the success of their organizational change initiatives, but will also better engage employees in change overall. This leads to creating a culture of continuous improvement within the organization. Readers will be able to develop their own strategy and plan for looking at change from a positive perspective—taking a project management approach to ensure structure behind change initiatives. Whether the reader leads large organizational change initiatives or is tasked with supporting organizational change initiatives, this book will provide a number of best practices and considerations to ensure success.

AUDIENCE

This book is intended for supervisors, managers, directors, senior leaders, and executives—anyone in a leadership role, including project managers and program managers as well as team leaders who manage projects within their organization. It is for change leaders, change managers, and change champions who are excited by the opportunities that change brings to the organization and who want to learn how to better manage change initiatives of any size and complexity. It will be of value to business managers, those in the C-suite, human resource professionals, business process improvement professionals, and consultants who manage change projects for their clients.

RESOURCES

A number of templates, surveys/questionnaires, assessments, and checklists are available from the Web Added Value™ Download Resource Center at www.jrosspub.com/wav. These tools and templates enable the reader to prepare for and engage employees in change within their own organization.

Throughout the book are a variety of best practice tips, examples, and mini case studies from my client work on change management, as well as a few stories shared by colleagues of mine. Each of the mini case studies provides a lesson for the reader.

Upon finishing this book, the reader will have a better understanding of the value of planning for organizational change and ensuring engagement of employees in change initiatives. The reader will have increased confidence and comfort in leading or participating in organizational change initiatives—from the simple to the most complex.

CHAPTER OVERVIEW

The chapters in this book take the reader from an introduction to organizational change, through to developing a strategy to enable creation of a culture of continuous change within the organization. The focus is significantly on the people-side of change in this book, as well as how to put structure around change so that it is well-planned and implemented. *Never* should the reader forget that change is *all about* engaging people so that they can adopt the change. By taking a more structured project management approach to launching and implementing organizational change, the organizational change leader is more likely to reach success. Successful change initiatives provide tremendous benefits to the organization and its employees. As that success increases from change initiative to change initiative, leaders will find that change becomes a bit easier in the organization, in that more employees are interested in change and willing to participate in the change—sharing feedback and ideas that enable more change to occur in the organization.

Chapter 1: Introduction to Leading Organizational Change

This chapter provides an introduction to change overall with a focus on the complexity of change today and the need for it to be viewed differently to increase the success of organizational change. This chapter provides the reader with an understanding of why change matters.

Chapter 2: Understanding About Change

This chapter provides background information on change to provide a base understanding of change within organizations. Factors that drive change and effects of change on the organization and individuals within the organization are shared. Understanding why organizations fail at change enables increasing the success of change overall. This chapter will set the stage for why focusing on employees is essential for organizational change success.

Chapter 3: Looking at Change from a Positive Perspective

This chapter will explore the value of looking at change from a positive viewpoint rather than the usual negative viewpoint. When change is seen as negative, it is usually because the change happens when an organization and its people are *forced* to change. Here, the value of looking at change from a positive and opportunity-driven perspective will be shared, as well as the importance of ensuring a vision for change. Best practices for communicating around change will start in this chapter.

Chapter 4: Building Change Capability Within the Organization

In this chapter, the focus will be on the need to build change capability within the organization. When this is done, it is easier to focus on change as a positive rather than a negative. Best practices for assessing the readiness for change as well as creating an environment for change will be shared. This chapter will also share the value of creating and sustaining a Change Center of Excellence (CoE) within the organization.

Chapter 5: The Value of Focusing on the People

The value of change from the perspective of the people—the employees of the organization—will be discussed in this chapter. Since people within the organization are the ones that control the success or failure of change, this chapter will focus on how to develop a mindset for change. This not only increases the success of change initiatives overall, but also enables for change initiatives that are not just supported, but frequently led, by staff level employees.

Chapter 6: Leading Change Across Cultural and Generational Boundaries

In leading change, the impact of cultural differences and generational boundaries cannot be ignored. This chapter provides information on how cultural differences and various generations in the workplace may impact change initiatives and shares the value of using cross-cultural and cross-generational teams to engage people in change.

Chapter 7: The Change Project

While this is not a book on project management, change initiatives are projects and should be managed as projects to increase their success. This chapter focuses on how to manage change initiatives taking a project management approach. Best practices will be shared for communicating with and engaging people in change from the beginning through to implementation. The use of change agents to help change *stick*, and the use of pilot groups to increase change adoption will be discussed.

Chapter 8: Continuous Communication and Engagement in Change

Communication can make or break a change initiative. This chapter dives deeper into the need for regular and continuous communication and engagement around change. Knowledge management as well as collaboration best practices will be discussed in this chapter.

Chapter 9: Continuous—But Not Chaotic!—Change

Organizations that recognize the value of change know that continuous change is a necessity in today's world. This chapter focuses on best practices for continuous change in the organization, relying on change teams to assist in keeping change moving forward.

Chapter 10: Getting Started

It isn't enough to just launch a change initiative; it must be planned. This chapter helps the reader get started by sharing several best practices and *things to do* in order to ensure that change becomes a regular part of the organization. The chapter helps readers who don't currently have a culture that supports continuous change begin to build such a culture within their own organization.

In summary, this book is meant to be a valuable resource for the reader, now and for years to come. By understanding the challenges in organizational change, how to overcome those challenges, how to plan for change, and how to engage employees in conversations around change, those involved in change initiatives will increase the success of those initiatives—driving profitability and revenue for their organization as well as attracting and retaining top talent. Regardless of roles and responsibilities within organizations, we *all must* get better about change, as change touches each of us. The more we understand about change, the more comfortable we become and the more we see change as a positive rather than a negative.

Happy Reading!

ACKNOWLEDGMENTS

First, I want to thank a very important person in my life—my husband, Yusuf Abudi. He has supported me in more ways than he can imagine. He helped to make this book a reality by allowing me to have the time to sit and write uninterrupted. He also supported this book by sharing a number of stories from his own experiences with his clients, as well as assisting with a number of the graphics throughout the book. He has been a great sounding board and remains my biggest fan—pushing me to accomplish much more than I often think I can.

Second, I would like to thank my mother, Fay Rusch Schmidt. She has supported me through all my endeavors throughout the years. She always reminds me that she is proud of who I have become and what I have accomplished. She has always been there for me and I can't imagine that I would be where I am without her. Her undying support, love, and encouragement has pushed me through many challenges.

Third, I want to thank my father-in-law, Khalil Abudi. He edited every chapter, providing an outsider's perspective that was invaluable.

Fourth, thank you also to a few of my colleagues and friends who contributed their own stories regarding change for this book: Mike Goehring, Paul Ross, and Francois Nadeau. Your time in putting down your stories in writing for me was greatly appreciated.

Fifth, a big thank you to Drew Gierman, my publisher at J. Ross Publishing. Without him this book would not be possible. And, last, but certainly not least, a thank you to all the editors and others at J. Ross Publishing who helped to get this book out the door.

Gina Abudi

ABOUT THE AUTHOR

Gina Abudi has over 25 years of experience in change management, project and process management, leadership development, and human resources. She is the President of Abudi Consulting Group, LLC, a management consulting firm serving mid-size to large global organizations.

As a consultant, Gina helps businesses develop and implement strategy around projects, processes, and people. A large percentage of her time is focused on organizational change initiatives. This includes efforts such as helping global organizations kick off large, complex change initiatives, setting up communication plans for change initiatives, helping to create and launch Change Management Centers of Excellence, and working with leadership to engage employees in change.

Ms. Abudi works closely with a variety of clients to develop and deliver customized workshops, seminars, and training programs to meet long-term strategic needs. She has been an adjunct faculty member at Granite State College in New Hampshire teaching in the masters of project management and masters of leadership graduate degree programs.

Gina is a professional speaker and keynote at many conferences, forums, and corporate and industry events on a variety of management, leadership, and project management topics. She is a professional member of the National Speakers Association (NSA). She is coauthor of *The Complete Idiot's Guide to Best Practices for Small Business* (Alpha Books, 2011) and contributing author of *Project Pain Reliever: A Just-In-Time Handbook for Anyone Managing Projects*, edited by Dave Garrett, CEO, projectmanagement.com and projectsatwork.com (J. Ross Publishing, 2011) and lead author of *Best Practices for Managing BPI Projects: Six Steps to Success* (J. Ross Publishing, 2015).

Gina served as President of the PMI® Massachusetts Bay Chapter Board of Directors and served on the Project Management Institute's Global Corporate Council as Chair of the Leadership Team. Gina has been honored as one of the Power 50 from PMI®—one of the 50 most influential executives in project management, working to move the profession forward. She currently works with the PMI Educational Foundation non-profit as a Community Engagement Member. Gina received her MBA from Simmons Graduate School of Management.

This book has free material available for download from the
Web Added Value™ resource center at *www.jrosspub.com*

At J. Ross Publishing we are committed to providing today's professional with practical, hands-on tools that enhance the learning experience and give readers an opportunity to apply what they have learned. That is why we offer free ancillary materials available for download on this book and all participating Web Added Value™ publications. These online resources may include interactive versions of material that appears in the book or supplemental templates, worksheets, models, plans, case studies, proposals, spreadsheets and assessment tools, among other things. Whenever you see the WAV™ symbol in any of our publications, it means bonus materials accompany the book and are available from the Web Added Value Download Resource Center at www.jrosspub.com.

Downloads for *Implementing Positive Organizational Change* include assessment tools, checklists, surveys/questionnaires, and templates that will enable readers to implement positive change within their own organizations.

1

INTRODUCTION TO LEADING ORGANIZATIONAL CHANGE

"Nothing is constant except change."
Heraclitus (ca. 513 B.C.E.), Greek philosopher

Change-savvy organizations are those organizations that recognize that change is a natural part of the organization. Leading change in organizations is a competency; one which executives and Boards of Directors look for in those they hire. This competency is a necessity whether the employee is an individual contributor or is in a leadership role.

Change-savvy organizations recognize:

- Change is a common occurrence
- Change is successful *only* when everyone in the organization is engaged in that change
- The vision for the change must be shared throughout the organization
- Communication around change must be frequent, honest, and thorough
- Change must be sold within the organization
- Culture is essential to the success of change efforts
- Change must be supported from the bottom of the organization
- Change requires a strategic project management approach

In a change-savvy organization, a typical change effort includes the following change team members:

- An executive who leads the change by providing the vision and ensuring understanding of how the change will benefit the organization *as well as* the individuals within the organization
- A project manager with experience in managing change projects and engaging the project team in change
- An individual who will be responsible for communications overall (this may be the project manager on smaller efforts)
- Team members who are engaged in the change and who represent a variety of groups or departments that will be affected by the change
- Managers who support the change by providing resources (team members) from their staff and ensuring that individuals with the right skills, experiences, and desire have the opportunity to work on the change initiative
- Individual contributors who are engaged in the change and are eager to undertake the challenge of change

Consider for a moment the best organizations you know or have read about; one thing they will have in common is the ability to adapt and change. To continue to move the organization forward, even in economic downturns and increased competition, change is necessary. Change—when done well—requires looking into the future and not just in the moment. When organizations look to the future, they are better positioned to adapt to what is happening around them. This enables them to be proactive and avoids being reactive. Reactive change rarely succeeds, but if it does, it usually has a tremendous impact on the resources of the organization and the customer base.

CHANGE VERSUS TRANSFORMATION—IS THERE A DIFFERENCE?

More frequently, organizations talk of going through a *transformation* as opposed to going through *change*.

> Merriam Webster defines change as: *to make or become different; to give a different course or direction to.* Transformation is defined as: *a complete or major change; an act of transforming or being transformed.*

Transformation is a dramatic shift in the culture of an organization. It impacts the strategy of the organization—all of its processes, procedures, and vision. It is a change in how the organization does business and addresses the needs of its customers. It is a change in the *beliefs* and *core values* of the organization. It changes business models and revenue models. Information technology (IT) systems change; performance systems change—the entire organization, effectively, changes. Transformation is an organization-wide effort and takes a significant period of time to accomplish. It may involve turning around a company in crisis or making a mediocre company a great company. Transformation requires significant planning. It is *always* a large and complex initiative. Transformational change alters nearly everything about the organization's culture.

> *Allison is a new Chief Executive Officer (CEO) in an organization that has been in business for six years. The organization has been in a start-up phase, but is ready to move to the next level. Allison, who has experience taking organizations from start-up to being more structured, will be responsible for launching a major transformation that will consist of restructuring the organization to include a number of business units with formal processes and leadership. This will entail hiring at least 50 new employees, in a variety of roles, over the next three to four years. It will also change IT systems in use as well as create structured performance management systems.*

Change, on the other hand, can be small and incremental, or it may be large and complex. Change may entail *tweaking* a process for paying vendor invoices so that the time to pay is reduced from 45 days to 30 days, or it may entail launching a new technology that enables better tracking of customer spend on products and services. Either effort requires a strategic project management approach to allow the change to be sustained.

> *One of my clients always looks at every change initiative as a significant change effort—he often uses the term, "transformation"—even if it is a change that will only impact one group. By looking at change as significant he knows that he will better plan for the effort—including ensuring early communications, engagement of employees, and spending more time thinking about the change and its value to the organization and its individuals. Every change he launches starts with sharing a vision for that change. He has come to realize that change requires changing the beliefs of some individuals within the organization.*

When we plan effectively for change—including thinking about the change and why it has to happen, who needs to be involved, and what a vision for the future will look like—we can make change successful whether it is a simple change impacting one group in the organization or a large, cross-functional complex change initiative.

THE COMPLEXITY OF CHANGE

Change is becoming increasingly complex, year after year. While in the past, organizations may have done just fine maintaining the status quo, that is becoming more difficult, if not impossible, these days. Increased competition, global economic impacts, securing and engaging top talent, increasing customer demands, price sensitivity, increased regulations, and many other factors all require companies to stay alert and continuously change and adapt to keep moving forward. This increased complexity requires a structured project management strategic approach to managing change.

> *"The rate of change is not going to slow down anytime soon. If anything, competition in most industries will probably speed up even more in the next few decades."* —John Kotter from *Leading Change*

Change is an emotional experience. It is no longer acceptable for leadership to demand that employees change. Rather, employees must be engaged in change for it to be successful. This, naturally, adds to the complexity of organizational change initiatives. If the employees of the organization are not engaged in the change, change cannot be sustained over the long term. Strategic project management looks at managing organizational change initiatives *beyond* the technical project management tasks to accomplish the project, along with focusing on engaging the organization in the change.

> *I have been working for one organization for over five years, specifically on helping them do a better job at launching change initiatives. Over time, this organization has seen the success ratio of their change initiatives improve by over 70% due to their effectiveness in getting support for change from the lowest levels in the organization and taking a strategic project management approach toward accomplishing the change initiatives.*

Change happens within an organization for one of two reasons:

- Mandated or forced: events occur that require the organization to respond, such as legal issues or regulatory changes

- Proactive: an organization wants to capitalize on something to improve the business, such as upcoming changes in the industry or marketplace

Too often, unfortunately, there is a tendency for organizations to be reactive rather than proactive—such as being forced to change because they are in a corner and feel they must fight their way out. Consider this situation:

> Joe is CEO of a public relations firm. For many years his company has been very strong in the industry with offices nationwide and some of the largest firms in the U.S. as their clients. No other competitor has ever really threatened them and, in fact, Joe and his team rarely think about competitors. Joe and his senior leadership team have been aware of a competitor that entered the market over three years ago, but saw no threat from them. After all, they were number one! Had Joe and his team been paying attention, however, they would have noticed that the competitor was gaining momentum year after year—expanding offices throughout the nation and retaining some big-name clients. One year, during a review of the business, it occurred to Joe that profits were decreasing and some of his organization's biggest clients were doing less and less work with his firm and, in fact, two of his biggest clients had terminated their contract. Sources indicated to Joe that these clients went with the competitor. Joe and his leadership team were now being forced to change in order to stay alive. That little competitor had grown up and was now a major threat to the business! Actions had to be taken and they had to be taken quickly.

In this situation, had Joe and his team paid attention, they would have been proactive; making changes as necessary to stay ahead of the competition. By waiting until they were losing clients, Joe and his team became reactive. Additionally, the complexity of change is also apparent in the type of change. Figure 1.1 depicts the levels of complexity common within organizations.

Increasing complexity requires earlier preparation as well as significantly more communication. According to a Change and Communication ROI Study conducted by Towers Watson in 2013–2014: *Organizations that change and communicate effectively are 3.5 times as likely to significantly outperform their peers.* McKinsey, in the July 2015 edition of *McKinsey Digital*, noted that research indicates that 70% of change initiatives fail to achieve

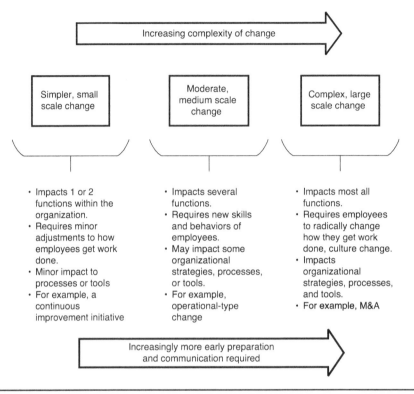

Increasing complexity of change

| Simpler, small scale change | Moderate, medium scale change | Complex, large scale change |

- Impacts 1 or 2 functions within the organization.
- Requires minor adjustments to how employees get work done.
- Minor impact to processes or tools
- For example, a continuous improvement initiative

- Impacts several functions.
- Requires new skills and behaviors of employees.
- May impact some organizational strategies, processes, or tools.
- For example, operational-type change

- Impacts most all functions.
- Requires employees to radically change how they get work done, culture change.
- Impacts organizational strategies, processes, and tools.
- For example, M&A

Increasingly more early preparation and communication required

Figure 1.1   Change complexity

their goals, primarily due to employee resistance and a lack of leadership support. This 70% failure rate has been fairly consistent over a number of decades. Organizations should be improving how they lead change initiatives—and, in particular, complex transformational change—but that does not appear to be happening.

Leading change requires a significant investment in communication and engagement of employees at all levels *throughout* the organization. Senior leadership alone cannot be a proponent of change; they must engage the employees—those individuals who accomplish the work of the organization—in change for it to be successful. Included in this book will be a number of best practices focusing on how to effectively communicate about change—from prior to launching the initiative, on through to implementing and checking in after implementation.

A good way to initiate change in the organization is by having a conversation with employees from across the organization—at all levels. This conversation is about the company and its future and is focused on how the employees are essential for the organization's survival. These preliminary conversations regarding change get the employees involved early-on in the process and inspire discussions about ways to improve how the work gets done. For one client, the conversation initially is before an all-hands meeting that starts with the CEO *telling attendees about his vision for the company—one, two, or three years out. This is all done before the client officially launches the change initiative and assigns a seasoned project manager to manage it.*

Utilizing a simple framework to lead organizational change efforts increases the chances that an organization will be successful in their efforts. Figure 1.2 is a simple framework for leading organizational change—from the very early stages of ensuring readiness and developing the change plan, on through to managing the change initiative and evaluating and maintaining the change to ensure it will stick. This framework will be followed throughout this book and will be flushed out more fully in Chapter 2.

Following this framework increases the chances of a successful organizational change effort because more employees are likely to adopt the change, which is essential to making the change *stick*. As can be seen in

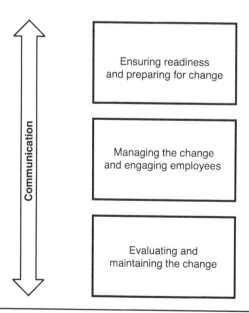

Figure 1.2   A framework for leading organizational change

Figure 1.2, communication is a constant. Through communication, organizational change leaders can better engage and get employees involved in the details of the change effort being undertaken within the organization.

One of my clients asked for advice on how to get employees engaged in the change initiative after it had already been launched. This particular client shared that the first communication that went out was to notify employees that the organization was in the second week of a major organizational change initiative and was interested in getting employees' thoughts on how the effort was progressing. As you can imagine, there were many angry and frustrated employees! Here the organization is asking for their thoughts around a change initiative that was already started and, in fact, many employees didn't even know the change initiative was happening—or why. From the perspective of employees, the organization didn't really want any response other than, "Looks good."

Such situations, as in the client example above, only serve to disengage employees even further from change and will create resisters to the change effort. Had the client followed the framework outlined in Figure 1.2, employees would have known *before* the change project was launched that change would be happening and why it needed to happen; and the organization would have been better prepared for involving and managing employees throughout that change.

WHY CHANGE MATTERS TODAY

The world is becoming a more complex place, which only increases the complexity of what organizations need to do to survive and thrive. Change matters today for any variety of reasons; but perhaps most importantly, it matters because *without* change, organizations *cannot* grow and prosper. Without change, employees within the organization cannot continue to develop professionally—the work becomes unexciting.

Without change, the strategy of the organization cannot be achieved. Change is a given for all organizations. Therefore, change is regular and constant in the most successful organizations; but that doesn't mean it needs to be chaotic. Chaotic change—or the perception of chaos—occurs when the organization has no strategy around change; change occurs just for the sake of change. There is no discipline around how changes are initiated in organizations where change is chaotic. In fact, in change-savvy organizations, change is a common occurrence often led by employees who regularly evaluate and improve how they get the work done.

One long-time client summed it up this way when sharing a variety of change initiatives that would be launched within the coming year, "Without change, there is no way our organization will survive over the long term. Our competitors are changing regularly; therefore we must also. We are not going to follow in their footsteps; rather, we will blaze our own path."

Employees Make the Difference!

While change is often process- or procedure-focused—essentially finding better ways of getting the work of the organization done to better meet client needs and compete effectively—there needs to be a behavioral component also. People within the organization must do something differently than they currently do. For example, if an organization changes a process for how data is entered into a customer relationship management system, this also requires people (those who enter the data) to behave differently—to change how they work to fit with the new process. The longer those people have been working with the old process (doing work the *old way*), the more difficult change becomes for them and more time must be invested to enable them to support and adapt to change. This is a simple fact of human nature.

Leading change requires an appreciation for change and the perceptions of those around the leader about the change. According to a survey conducted by PeopleNRG, Inc., of 907 Human Resources, Organizational Development, and workplace learning professionals and leaders in the United States in 2011, over 80% of employees accept change when that change is supported by influential *nonleaders* in the organization—those are your employees! While certainly change cannot happen without top-down support, change is certainly not possible at all without employees' participation from throughout the organization.

> *"People don't resist change. They resist being changed!"*—Peter Senge

Figure 1.3 provides an example of how one organization engaged employees in change from the start. Once engaged, employees then owned the change initiative launched from the executive level.

In this particular case, a retail organization changed leadership at the top. The CEO and other top executives had experience with engaging employees in change and knew the value in doing so. As can be seen in Figure 1.3, once the original vision of the change that is being proposed is shared with employees, leadership requests feedback. For this client, regarding one particular complex change initiative, feedback was provided over a

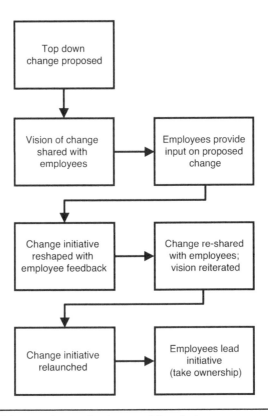

Figure 1.3   Involving employees in change at the start

three-week time period. Employees had a variety of options to provide that feedback, including:

- Focus group meetings
- Department meetings
- All-staff meetings
- An internal website specifically launched to support the change
- Directly to anyone on the leadership team via the group's *open door* policy
- An online survey sent to all staff requesting feedback

Employees provided significant thoughts, suggestions, ideas, and concerns about the initial proposal for the change initiative as well as the vision. Through the employees' insight and contributions, the executive leadership team was able to reframe the change initiative incorporating many of the ideas of the employees, as well as to address many concerns raised. The result was a change initiative that was, effectively, shaped by employees. Because employees were involved in the process of framing the change initiative early on, they were *sold* on the initiative. Employees were engaged in the initiative and, in this situation, when leadership asked for employees to lead the change, there was no delay in getting many employees to commit to taking charge of the project. They owned it.

Let's discuss engaging employees in change by looking at change from a three-stage perspective as can be seen in Figure 1.4.

In Figure 1.4, the organization reviews how the work is done today. This effort *requires* engaging employees because, let's be honest, the work of the organization is done by the employees. In the transition stage, the organization must ensure understanding of how the organization needs to change—which translates to: how does the work of the individual employee need to change to support the organizational change. The last stage is the future: how the work is *then* done in the organization—the organization of tomorrow.

When organizations engage employees in change initiatives early—*before* the change is finalized and officially rolled out—they are more apt to sell them on the change. Contrast the previous situation with this one:

The new CEO of an engineering firm was not used to engaging staff in change initiatives. At the last organization he led, he simply let employees know that change was happening and when it would be effective. At this firm, however, the employees expected to be engaged. The first time the CEO attempted to launch a change initiative, he learned quickly that without engaging the employees, it would not be successful. The first initiative that was a process change was considered a failure upon implementation, as less than 10% of the employees used the new process. It wasn't a matter of refusing to adapt but rather, for a significant number of employees,

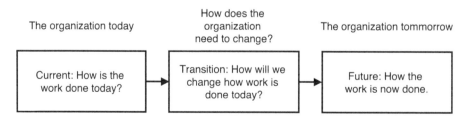

Figure 1.4   Stages of change—today to tomorrow

the new process simply did not work, given limitations around technology and resources within the organization.

In this example, had the CEO engaged those employees who were working with the current process on how to change the process to increase effectiveness and efficiency, he would have learned that other changes had to occur *before* changing the current process.

A 2013 study by the Katzenbach Center found three major obstacles to implementing organizational change initiatives. One of those obstacles is that transformational change is often decided upon, planned by, and implemented by the executives in an organization with little or no input from the employees. This not only removes the possibility of getting input necessary for a successful change from employees, but also reduces front-line ownership of the change. In the Katzenbach Center survey, *44% of participants reported not understanding the changes they were expected to make, and 38% said they didn't agree with the change.* When employees are not involved in the change, they will not support it, which leads to failed change initiatives. Failed change initiatives are costly, not just to the bottom line but they also disengage employees from the workplace through decreased morale and a feeling of failure. No one wants to work for an organization that is considered a failure.

Throughout this book, a number of ways to engage employees in change will be shared. *Every* successful organizational change initiative has been successful *because* employees were engaged in the effort. If an organization does not engage employees in change, the change is simply *not* successful. This cannot be stated enough!

CHANGE CAN BE EXCITING AND FUN!

So far we have talked about the challenges and complexity of change. Reading up to this point, an organization may think change has to be the worst thing to do. No! Change, when positioned well, is actually exciting and fun. It opens up the possibility for creation of something new—improvement that initiates increased success, the ability to learn new skills and build knowledge, and many other achievements for both the organization and the individuals within the organization. If the organization looks at change as a regular occurrence within the organization, change is perceived as less an emotional experience and more an expectation—but only when handled well.

Organizations that plan for change—through early and frequent communication, sharing the vision, ensuring cross-functional participation, and rewarding for changing—are more likely to have employees who see the opportunity and the exciting side of change. Those organizations that look at change in the same way as Joe and his executive team of the public relations company looked at change in the earlier example will find themselves surrounded by employees who are apathetic and view change as a major inconvenience. These are the organizations that will not survive over the long term. Employees have low morale and feel disappointment—in themselves and in their organization. The best employees—top talent—will leave for the competition.

In Chapter 5 the topic of change agents will be discussed. These are the individuals who embrace change—change is exciting for these employees. Leadership alone cannot make change fun and exciting, but employees who are engaged in change—*who champion change*—will display an excitement around change that will be contagious to those around them.

If organizations position change as an opportunity—the chance to do something wonderful and unique—change can be seen as exciting and fun for employees. This will be the case even when the change is a large undertaking within the organization.

The balance of the chapters will give insight into answering key questions when discussing change with employees. Table 1.1 provides a sampling of questions that those who are initiating change should be prepared to answer during initial conversations with employees.

These questions answer the *who, what, where, when, why,* and *how* of change.

Table 1.1   Questions to be answered about change

Questions Around Change
Place a checkmark next to the questions you are prepared to answer; determine answers to those you cannot answer.
Why does the change have to happen?
What is driving the change?
What is the vision for the change?
What is in it for me?
Who will be involved?
When will it start?
Where will it occur (what department, division, etc.)?
How will it look when done?
Who is impacted?
Who needs to be involved?

2

UNDERSTANDING ABOUT CHANGE

"Nothing is so painful to the human mind
as a great and sudden change."
Mary Shelley, *Frankenstein*

Understanding about change seems easy enough; but there is much involved in truly *understanding change*—its impact on the organization, business units and divisions, departments, workgroups, as well as the individual employees. The more that is understood about the change, the more likely leaders can launch change initiatives that:

- Make sense for the organization
- Are tied to the vision of the organization
- Have the right people involved on the initiative
- Are communicated about regularly and sufficiently
- Are successful upon implementation
- Are retained over time
- Have a positive impact on the bottom line

I frequently work with clients to look at why change didn't stick over the long term. In most cases, the change didn't stick because employees were never really involved in the change from the beginning and therefore never supported the change or bought into it overall. Additionally, change is not looked at as a strategic initiative

*and therefore a strategic project management approach is not uti-
lized in the organization.*

Understanding about change is essential in planning for and managing the
change effectively. It is also necessary to answer the questions (see Table
1.1) regarding the change that will inevitably arise when leaders introduce
the need for change to employees.

Understanding about change means that leaders are better prepared to
initiate change in the organization, using a strategic project management
approach, and taking the steps necessary to ensure buy-in and commitment to the change from throughout the organization. Understanding
about change means that leaders have an appreciation for the complexity
of change, and that includes the perception of change by the employees.
They understand that what is considered an *earth-shattering* change to one
employee may be considered a *no-big-deal* change to another.

Understanding the fact that change must be sold within the organization requires skills in:

- Communication
- Listening
- Influencing
- Empathy
- Building relationships

Understanding about change means that leaders acknowledge that change
does not come easy. From the very beginning, when discussing change with
the employees, it is essential to instill a sense of urgency about the change
and why it has to happen. As the change initiative moves along, especially
for longer change initiatives, that sense of urgency has to be shared again
and again.

> *Jeremy couldn't understand why, after he spent two hours in an
> all-staff meeting discussing the change initiative and why it had to
> happen, it seemed that, only three weeks later, he needed to share
> the "why" of the change again. Weren't people paying attention?
> Did they forget? How many times would he have to repeat why this
> change was happening?*

Jeremy will need to share the *why* of the change more than just once at
the beginning of launching the change initiative—this will drive a sense of
urgency about the change. That sense of urgency will need to be instilled

throughout the initiative to ensure continued understanding (*why are we doing this*) and continued support (*we agree it has to happen*) for the change. This chapter will explore a number of key aspects of change such as factors that drive change, effects of change on the organization and the individuals, and identifying and overcoming obstacles to change. An understanding of these key aspects of change enables more effective upfront planning and engagement of the right resources on the change initiative.

A CHANGE MANAGEMENT FRAMEWORK

Chapter 1 briefly introduced a change management framework (see Figure 1.2). Consider this framework a high-level overview of how change might be launched and managed in order to choose a path to greater success of the initiative, regardless of its complexity. Table 2.1 looks at the framework in more detail.

In Table 2.1, a number of questions are to be considered as well as what needs to be done in each stage in order to ensure the change initiative will be successful and last over time. The more prepared an organization is for the change initiative—from considering if it is the right time to launch the change through to determining how to ensure the change will *stick*—the more likely that the change initiative will engage employees throughout the organization and be successful. A successful change initiative has the following characteristics:

- Employees are engaged and support and champion the change
- Early on in the change initiative, employees from throughout the organization understand and embrace why the change has to happen, that is, employees *see* the vision for the change
- A project manager is engaged in leading the change project
- The organization takes a strategic project management approach to change
- Communications begin early on in the change initiative and continue throughout
- Feedback is requested from all levels of employees throughout the change initiative
- Early on in the initiative, employees understand how they will be prepared for working with the change—whether that is by being trained, having a transition period, or in some other way.

All of this requires *up-front* planning of the change initiative. Following a change management framework, such as what is described in Table 2.1,

Table 2.1   Accomplishing components of the framework for leading organizational change

To accomplish this...	Consider this...	And do this...
Ensure readiness and prepare for change	• Is it the right time to initiate this change? (What else is going on in the organization?) • What is the vision of the change? • Does the change support the organization's strategy? • Are there resources available to work on this change? • Have past change initiatives gone well? If not, what have we learned? • What processes and procedures must change? • What skills and knowledge are necessary to support the change after implementation? • Does the organizational structure need to change? • How complex is the change? • Is there leadership sponsorship for the change?	• Share the vision for the change. • Share the "why" of the change. • Share the benefit of the change to the organization and to the employees. • Ask employees for input about the change before it is finalized. • Communicate initially in a variety of ways to reach and engage the largest group of employees. • If past initiatives have not gone well, be prepared to communicate what will be different this time.
Manage the change and engage employees	• Who needs to lead the change initiative? • Who comprises the team working on the change? Is cross-functional representation necessary? • What are the roles and responsibilities of those working on the change? • Who are champions of the change, and who might resist the change? • What will regular, ongoing communications look like, and how will they be delivered?	• Develop a project plan for the change initiative. • Develop a communication plan for the change initiative to ensure and enable employee engagement. • Create a stakeholder support group to help in communications and sharing information. • Develop a stakeholder list— identify champions, resisters, and those who appear indifferent to the change.

To accomplish this...	Consider this...	And do this...
Evaluate and maintain the change	• What is the right time to ask for feedback about the initiative? • What has been done in order to ensure the change will "stick?" • What is the mechanism for tracking what is working on the change initiative and what is not working? • Have formal adjustments been made to accommodate the change (e.g., reporting structures) as well as looking at the impact on informal components (e.g., internal networks)?	• Early on in launching the change initiative, determine how feedback will be gathered and acted upon. • Regularly ask for feedback on how the change is progressing through a variety of channels. • Ensure plans for training, transitioning, follow up, and whatever else is necessary to ensure the change "sticks." • Review adjustments that have been made to ensure the change is usable and sticks (performance feedback, systems updated, documentation updated, IT systems updated, etc.).

facilitates effective planning of the change initiative *before* it is actually launched within the organization.

As an activity in launching change initiatives, I ask client employees for their perceptions of change, based on past experiences. Responses are captured on flip charts. Responses gathered from a recent client change are shown in Table 2.2.

When asked at "Client A" how these words come to mind, employees who participated in the activity noted the following reasons:

- *Past experiences at this company and also at other employers*
- *Lack of communication from leadership about the change and why it was happening*
- *Inability to provide feedback, thoughts, or ideas about the change*
- *Feeling that people would lose their jobs even though leadership told them "no"*

This effort enables me to understand the perception that employees will likely have around the current initiative. This knowledge

Table 2.2 Client A: positive and negative perceptions of change

Positive Perceptions	Negative Perceptions
• Opportunity	• Anger
• Efficiencies	• Unnecessary
• New ideas	• Fear
• Innovation	• Waste of time
• Better service	• People getting hurt
• Better products	• Confusion
• Happier customers	• Frustration
• Happy employees	• Low morale
• Improvements	• Lost jobs
• Out of a rut	• Unhappy employees
• Excitement	• Worry
• Thrilling	• Mistrust

enables more effective planning on communications and engagement of employees.

What Kind of Change Is This?

An understanding of the type of change faced in the organization helps everyone to realize the impact of that change and the amount of up-front planning and communication that needs to be done. It also helps with recognizing who should be involved in the change effort. Table 2.3 describes three types of change within the organization.

The more complex the change—such as a transformational change as described in Table 2.3—the more time that needs to be spent up front in designing the change and engaging employees from throughout the organization. Recall that in Chapter 1, Figure 1.1 looked at change complexity from a similar perspective. *Complexity* and *type* of change are intertwined.

One of my clients merged with another organization and, due to the merger, would be making significant changes in the organization. These changes would impact the organization's vision, the clients they served, technology used, the services they offered, as well as the roles and responsibilities of employees. This was a transformational change for the organization. Once the merger was confirmed, leadership reached out to all employees—through an all-hands virtual meeting, through e-mail from the Chief Executive Officer (CEO), as

Table 2.3 Types of change

Type of Change	When the change...
Transformational	• Is large scale and impacts most functions in the organization • Requires employees to radically change how they work • Impacts organizational structures, strategies, processes, and/or tools
Operational or Incremental	• Is moderate in scale and impacts several functions • Requires employees to learn new skills and behaviors but retains basic responsibilities and workloads • May impact some organizational strategies, processes, and/or tools
Continuous Improvement	• Is small scale and impacts only a few functions • Requires minor changes to how employees do their roles, but there is no change in skills, responsibilities, or workloads • Minor impact to processes and/or tools

well as through small group meetings with various executives of the organization—to explain what was happening and why, the impact on employees, and how the organization needed employees to help them in this initiative.

Figure 2.1 provides some high-level considerations prior to formally initiating each type of change.

The more complex the change, the more time that should be spent up front in preparing for the change, *prior* to getting the entire organization involved in the change. As an example, the executive leadership team may begin to consider impacts of transformational changes three plus months *prior* to actually sharing information with the organization as a whole.

Consider the example provided earlier of my client involved in a merger. Approximately six months prior to formally confirming the merger, the executives considered the impact of the merger on the following:

- *Overall organizational structure, including reporting relationships*
- *Performance management*
- *Incentives and reward structures in place*
- *Benefits for employees*
- *Client and market base*

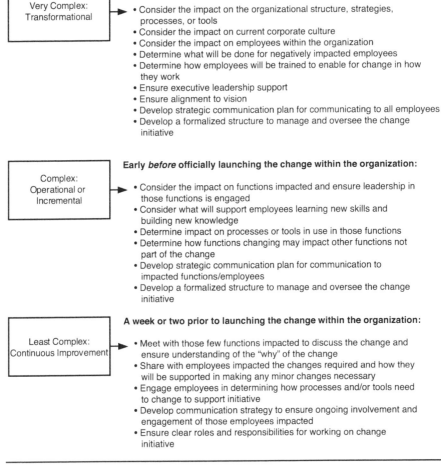

Figure 2.1   Preliminary considerations to kick off each type of change

Executives also considered:

- *Necessary roles and responsibilities under the new organizational structure*
- *Notification to employees negatively impacted by the merger*
- *How to get employees involved in supporting the merger*
- *Communication of the value of the merger for the organization's long-term viability as well as the value to employees*

Those primarily involved in these earlier stages of preparing was the executive leadership team. However, the executive leadership team also knew that involvement of leaders from other areas of the organization was necessary in order to ensure complete understanding of various impacts throughout the organization. The main goal of preparing so early on was to make certain that the organization would not be thrown into turmoil, which would negatively impact work with clients.

Transformational changes impact the culture of the organization. This is a given. Impact on the organizational culture requires changes to the vision, mission, and core values, expectations of employees and leaders, norms, standards, and processes, as well as the working habits of employees. In many cases, it may be seen as a rebirth of the organization.

Regardless of the level of complexity, it is essential that the employees in the business are kept in mind. Even if the impact of the change may seem minor on the employees, it is critical to engage employees in the change process and acknowledge perceptions of the change. Figure 2.2 provides a partial communication overview plan with a specific focus on engaging employees in the change.

Sender	Overall Timing (project phase)	Audience	Primary Focus	Message Content	Delivery Method	Date
CEO	Defining Project	All Employees	Announce upcoming change	Vision for change Explanation of change What to expect	All-hands meeting (virtual) Email follow up	January 8
Change Leader (Sponsor)	Defining Project	Identified change team members	Team responsibilities and team building	High level overview of change project Roles and responsibilities for team members Estimated timeline Team building activities Determine how team members will work together	Team meeting (face-to-face)	January 15
Change Leader (Sponsor)	Project Start	Potential (identified) Stakeholder Support Committee members	Role and responsibilities of Stakeholder Support Committee	Role and responsibilities of Stakeholder Support Committee members Benefit of being on committee Commitment expectations	Email	February 15

Figure 2.2 Partially completed communication plan for engaging employees

By outlining communications and developing the communication plan, the organization is more likely to engage employees early on and throughout the change. As the change initiative progresses, often communication falls off and diminishes in amount and quality. As problems arise, as conflicts occur, as timelines loom near—communications are the last thing on any leader's mind!

The communication plan shown in Figure 2.2 includes:

- The name of the sender of the communication piece
- The overall timing of when the communication is due to be sent
- The audience who will receive the communication
- The primary focus of the communication (its purpose)
- The message content (bulleted items of what should be included in the communication)
- The delivery method to be used (e-mails, meeting, phone call, internal website, etc.)
- The date the communication is being sent

For larger, complex organizational change initiatives, assign individuals who are to be responsible for communications overall. For example, looking at Figure 2.2, once the Stakeholder Support Committee is selected and in place, the communication responsibilities may become the role of that group, organization-wide. This information would be included in the communication strategy and would designate the *sender* as the Stakeholder Support Committee. The delivery method may be *casual conversations* or *department meetings.*

> *I have heard of situations where so many changes were happening at once that the organization was in complete disarray. For example, one organization was introducing a new product line while in the middle of an expansion and relocation, and undertaking a major technology upgrade! WOW! To top it all off, the organization had not really planned for any of the changes and, in fact, in this case, changes were launched without an understanding of what was happening elsewhere in the organization—a communication failure right from the start. The group releasing the new product line did not realize that a major technology upgrade was happening at the same time. And while there was knowledge of relocation and expansion, neither group knew it was happening at the same time.*

THE PROCESS OF CHANGE AND STAGES OF CHANGE

Having a process around how change is managed within the organization will expedite increased consistency and effectiveness when change initiatives are launched, as well as enabling individuals to move through the stages of adapting to change (see Figure 2.3). First, let's discuss the stages of change with a focus on how leadership can assist in moving individuals through the stages from fear and resistance to accepting and embracing change.

While Figure 2.3 is certainly not representative of *every* employee in the organization, it is representative of many employees. Change worries people. It is concerning. It requires us to change how we work, as well as our behavior. People prefer the status quo—they are used to it and it is comfortable. Change requires us to move *outside* our comfort zone, which often causes resistance. Leaders who recognize this about change and employees are more able to get employees comfortable with change earlier on in the process.

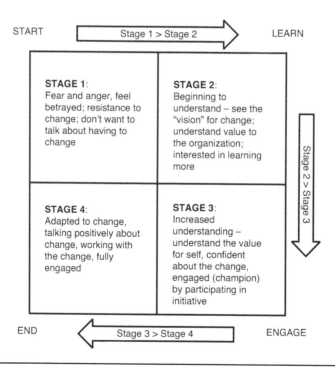

Figure 2.3   Stages of adapting to change for individuals

In Figure 2.3, as soon as they hear of a change—whether through informal or formal channels—employees begin (Stage 1) with being apprehensive, angry, and worried about change. They resist the change; they don't want to talk about the need for change. Leadership needs to recognize that this is a common occurrence. The responsibility lies with leaders to ensure an understanding around change by sharing the vision, creating an urgency for change, and engaging the employees to learn more (Stage 2).

Eventually, the comfort level with the change begins to increase (Stage 3). Employees have learned even more about the change and see the value not just for the organization, but for them personally. Leadership has enabled this increased learning and understanding, as well as acceptance of the upcoming change, by continuing conversations around the change, and enabling employees to participate in shaping the change through providing their opinions, ideas, and suggestions. Leadership should address concerns with the change through a variety of channels, such as via e-mails, internal sites, or focus groups.

A client of mine, the CEO of a pharmaceutical company, does an excellent job in engaging his employees in change by taking the time to address every concern he receives. Over time, he has learned what those concerns are likely to be and uses that knowledge to address concerns in his initial discussions with employees about the upcoming change. In this way, he has effectively sped up the process of moving through the stages and engages employees much sooner (Stage 3) than he has in the past.

It should be noted that, as much as it would make life easier, it is not likely that *every* employee will become a champion of the change initiative. More on champions and resisters to change will be covered in Chapter 5.

In Stage 4, employees have adapted to change. This doesn't mean the conversation ends. As discussed earlier, communications around change should continue even *after* implementation of the change. This facilitates continued engagement and ensures that the change will *stick*. If leadership implements change and then just makes the assumption that it is done, they are more likely to see employees revert back to the old way of working. This doesn't mean to imply that we must forever continue to *check in* and communicate about the change that was implemented. Depending on the complexity of the change, and the number of individuals within the organization impacted, a plan for continued communications for a period of time is wise. Additionally, this continued communication helps to

Table 2.4   Continuing conversations after implementation

Type (Complexity) of Change	Continue Conversations for...
Continuous Improvement (Small Scale)	Minimum of 2-3 weeks after implementation
Operational or Incremental (Moderate)	Minimum of 2-3 months after implementation
Transformational (Complex)	Minimum of 6-9 months after implementation

recognize areas where the change may not be working and adjustments must be made.

While there is no magic number as to how long to continue conversations around change *after* implementation, Table 2.4 provides some guidelines for doing so based on the type of change the organization is implementing.

These rules for continuing conversations to ensure change will stick are not *hard-and-fast.* Beyond the complexity of the change, other factors that may change the time spent in conversations about the change after implementation include:

- The percentage of employees who have adopted the change
- The amount of input employees had in defining the change early on
- How engaged employees were early on in the initiative
- The understanding of the value of the change for the employee

Figure 2.4 provides a four-step process for engaging others in change.

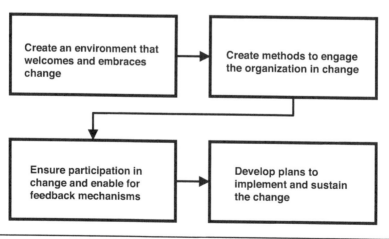

Figure 2.4   A simple four-step process to engage others in change

Table 2.5   Brief details of each step in the four-step process

Step	What Happens in this Step
Create an environment that welcomes and embraces change	• Ensure a vision for the change and share that vision • Communicate the value of the change—why it's important—use a variety of communication channels • Share benefits for the organization as well as for the individuals • Enable for discussions around change through focus groups or online forums • Acknowledge past failed change initiatives and discuss lessons learned
Create methods to engage the organization in change	• Review past communication efforts—What worked? What did not work? • Determine champions, resisters, and those who may be indifferent • Develop a communication strategy that utilizes a variety of channels and methods for communicating on the change initiative • Develop a Stakeholder Support Committee group comprised of members from throughout the organization • Determine "change agents" to deploy throughout the organization
Ensure participation in change and enable for feedback mechanism	• Develop regular and a variety of methods for providing feedback before, during, and after implementation of the change (include anonymous ways to provide feedback) • Ensure a charter for the Stakeholder Support Committee as well as a forum to bring feedback back to the change leadership team • Ensure feedback requests as part of the communication strategy through surveys and focus group forums
Develop plans to implement and sustain the change	• Develop an overall project plan for the change initiative (incorporate communication plans, training plans, pilot testing plans, stakeholder lists, and other relevant documents) • Determine the impact on formal as well as informal structures within the organization and develop a plan to ensure that structures that are impacted will be revised accordingly • Develop a plan to ensure the change—once implemented—is sustained such as through follow-up training sessions, checking in formally and informally with those impacted, additional feedback sessions, surveys, etc.

This is one of many possible steps that an organization may have in place to engage in change and increase the success of the change initiative, regardless of its complexity. Table 2.5 provides more details on each step.

To enable a more comprehensive initiative overall, think about the organizational change from the early stages (before actually launching the change initiative) through to implementation and then follow up with those impacted after *go live* (refer back to Table 2.4).

Table 2.5 provides considerations for each step in the process—all with a focus on engaging others in the change initiative.

> *One of my clients launches each change initiative—regardless of its complexity—by engaging employees from throughout the organization early on to discuss the change and get their perceptions about the change. This enables the client to not only begin to sell the change to his employees, but also to gauge how successful the change will be overall. As he perceives more resistance, he realizes he needs to spend more time understanding why the particular change is not garnering support. He continues conversations with employees, helping them to frame the change in ways that work for them and also allows him to meet his goals.*

While evaluating past change initiatives that had failed within a global organization, the questions depicted in Figure 2.5 were asked of the client's employees via an online survey and followed up in small virtual focus group sessions.

In addition to questions regarding past change initiatives, additional open-ended questions were asked to help in framing communications and conversations around change. Gathering this data and addressing perceptions of change within the organization has enabled the movement of employees through the stages of changes more quickly. This has been done through acknowledging past failed initiatives and outlining for employees how the next change initiative will be different. In other words, sharing the lessons learned from the failed change initiative.

> *For another client, the question—"If you were leading a change initiative, what would you do to help employees adopt the change and be comfortable with it?"—was used not only for communication planning, but also to engage employees in participating in various roles based on their ideas and suggestions. For example, one employee, in her response, suggested that as a remote worker, she often felt left out of change initiatives and would highly recommend that*

Statement...	Strongly Disagree				Strongly Agree
	1	2	3	4	5
The vision for the change was clearly communicated and understood.					
I understood not only the impact of the change on the organization, but also how the change would impact me personally.					
I felt that I had a hand in shaping the change initiative.					
Communications about the change were effective and sufficient.					
I knew early on what I would need to do to be effective in working with the change and how the organization would support me.					
I felt like everything that needed to be considered regarding the change was considered early on and with the input of employees.					
Statement...	**Open response...**				
Considering the last change initiative in which you were involved within the organization, what would have increased your comfort, or made you even more supportive, of the change?					
What worries or concerns you about change within the organization? (Be specific.)					
If you were leading a change initiative, what would you do to help employees adopt the change and be comfortable with it?					

Figure 2.5   Survey sent to employees to evaluate past failed change initiatives

all remote employees be brought to the headquarters for a day, in order to not only share their perceptions of the change, but also to help frame the change. As a bonus, she noted that it would be a great opportunity to get remote employees together. The VP of Finance (the individual leading the change) took her up on her idea and asked her to draft the agenda for the day.

When organizations are up against past failed change initiatives, consider the following ways to build momentum for a new change initiative:

- Look for a past successful change, even if a smaller one
- Find a small sign of how change is already happening within the organization and highlight it to all employees
- Begin communicating much earlier and more often than is usually done within the organization

Embracing the fact that change is worrisome for many employees will facilitate the crafting of a better, more robust communication plan—one that decreases the fear of change and engages employees in crafting the final product of that change.

STRATEGIC INITIATIVES AND ORGANIZATIONAL CHANGE

Strategic initiatives launched within an organization are often significant change initiatives and must be considered from the perspective of change. Let's look at an example.

An organization has completed their strategic plan for the year. They have determined that the following two key initiatives are to be a focus of the organization over the next two to five years:

- Increase market share (growth in international markets, new product development, new marketing channels within the U.S., evaluation of partner channels)
- Improve time-to-market (improved processes in product identification through development, reduction in manufacturing sites to reduce costs)

These strategic initiatives are directly tied to the bottom line and increased profitability within the organization. Let's focus on just one of these to continue our example—*improved time-to-market*. This strategic goal requires two focus areas: improved processes as well as a reduction in manufacturing sites. Both of these focus areas are significant change initiatives. The first—improved processes—will require employees to change how they get the work done currently. The second—reduction in manufacturing sites—will cause job loss, which often causes a decrease in employee morale. In order to be successful with these strategic initiatives, the organization will need to consider whether or not they are ready for the change that is coming. Figure 2.6 provides a checklist to gauge the readiness for

Yes/No	Is the organization ready for change...
	Is the vision for the change clear and understandable to employees?
	Do the employees understand the need for change?
	Does leadership understand the impact of the change on employees?
	Is success of the end result clearly defined? (Do we know what success looks like?)
	Are there clear metrics for measuring the success of the change?
	Does senior leadership support the change?
	Are there resources available to implement this change?

Figure 2.6   Is the organization ready for change?

launching a specific strategic and impactful change within the organization.

Figure 2.6 includes a partial listing of questions that should be considered to determine if the organization is ready for the change. A more complete listing of questions can be found as a downloadable file from the Web Added Value™ Resource Center at www.jrosspub.com/wav. If leadership is unable to respond *yes* to each of the questions posted in Figure 2.6, the change initiative should not move forward and needs to be reexamined.

Let's go back to our example. In this situation, after further evaluation, leadership determined that an improvement in processes may actually negate the need to reduce manufacturing sites immediately. Additionally, there were not sufficient resources to focus on both improving processes *and* reducing manufacturing sites. It was also apparent through initial conversations with managers that employees would more easily support improving processes over shutting down manufacturing sites, for obvious reasons.

Given that strategic initiatives often indicate significant and transformational change, organizations will do well to reduce the number of strategic initiatives to be accomplished in any given year. Too often, organizations launch too many strategic initiatives without understanding that these initiatives are significant change initiatives. While there is no magic number, as a best practice two or three strategic initiatives in any given year are probably sufficient and more easily able to be managed within an organization. Of course, a smaller organization may only be able to accomplish one or two strategic initiatives. In making a decision as to which strategic initiatives to launch, an organization must consider:

- Available resources
- Available budget
- Tolerance for change
- How complex the change will be for the organization
- How supportive leadership is of the initiative

FACTORS THAT DRIVE CHANGE

Organizations don't launch change initiatives simply for the fun of doing so. Organizational change is driven by any number of internal and external factors. Understanding the factors that are driving change within the organization will enable better conversations with employees as well as acceptance of the change. Too often, organizations consider the desire for increased profitability or more revenue or a reduction in costs as the need for change. While certainly those are all good and valid reasons to launch change, there are many other factors that drive the need for substantial and transformative change within the organization, and they must be considered. As part of an overall review of the change initiative being launched within the organization, consider both internal and external factors that are driving that change to occur. Understanding these factors will bring about a better overall change management plan, as well as a communication plan to share and gain support for the change from throughout the organization. Table 2.6 provides a listing of both internal and external factors that drive organizational change initiatives.

Organizational leaders who are wise are able to speak to factors that may at some point impact their organization. This initiates being proactive

Table 2.6   Internal and external factors that drive organizational change

Internal Factors	External Factors
• Mergers and acquisition	• New competition in the market
• Changes in senior leadership or other management	• Industry changes
• New Board of Directors	• Changing customer demands
• Change in organizational vision	• Cost pressures
• Employee dissatisfaction	• Technology
• Organizational growth	• Regulatory changes
• Business unit or department mergers	• Economic changes
• Performance failures	• Market niche
	• Vendor or supplier changes

rather than having to react to a factor that is impacting the organization and its longevity.

> *One pharmaceutical company neglected to keep an eye on the competition. A competitor that was struggling was ignored. About two years after losing track of the competitor, the competitor suddenly popped back up on their radar. The competitor had moved ahead of them and had taken a significant amount of market share. Panic struck the organization as they struggled to determine how to stay alive in what was now a competitive situation.*

Had the organization kept track of the competitor, they would have been proactive in staying ahead of the competition and could have kept the competitor in their rear-view mirror. Instead, they panicked and reacted to what they perceived to be a competitor who rose up rapidly. In reality, the competitor did not move up rapidly, it made slow and steady progress to become competitive. In reacting to the competitor, layoffs occurred in the organization, and they lost significant market share. As of the writing of this chapter, there were talks of the organization closing their doors.

EFFECTS OF CHANGE ON THE ORGANIZATION

Change has tremendous effects on the organization overall. When proactive, change tends to have a positive effect on the organization. It encourages growth, prosperity, long-term sustainability and helps the organization to meet customer needs as well as attract new customers to the business. It enables the organization to attract and retain top talent. It promotes more effective competition as well as reducing time-to-market for products and services. And, of equal importance, it brings about a positive impact on the organizational culture overall. Table 2.7 shows some key areas for change within the organization.

Table 2.7   Key areas for change in the organization

• Improved customer relationships	• Processes, policies, and procedures
• Marketplace demands	• Introduction of new products and services
• Increased competition	• New technologies
• Increased market share	• Use of technology to enable employee performance
• Industry changes	
• Mergers and acquisitions	

Many of these key areas for change are also factors that drive change within an organization. The most successful organizations are those that see change as a constant in their organization rather than to simply fix a problem. Constant, however, should not mean chaotic. Constant, proactive change is equal to innovation within the organization and to implementation of new ideas. Consider this example:

> *All Company Training wants to launch a strategic change initiative to increase innovation in product creation, thereby getting new and unique training products out to the market more quickly than their competitors. Their vision around this initiative is to strengthen the organization and increase their market share in an industry that is very competitive. Given the strategic nature of this initiative and the fact that it will require significant cash and resource investment, All Company Training will begin the initiative by working with those employees closest to the customer to understand more about the customer's current needs and future desires. Additionally, they will also engage employees by asking them to help in researching trends and new technologies and by soliciting their ideas on new, innovative products. Additionally, All Company Training regularly asks employees for their thoughts concerning new product ideas. This information will be compiled and evaluated now that All Company Training is ready to launch this strategic change initiative.*

Table 2.8   Why change efforts fail in organizations

• Lack of a shared vision, or lack of communication around a vision	• The belief that "if it ain't broke, don't fix it" (status quo)
• No repercussions for doing work "the old way"	• So much going on that the organization cannot manage one more thing
• No support/buy-in for the change from leadership	• Lack of understanding around the change and its purpose
• Lack of honest and open communications about the change and its impact	• Distrust of leadership
• No input from employees/lack of engaging employees in the change	• Significant risk of the change and not addressing/managing the risk
• Ignoring feedback or input from employees	• Not enabling employees the time and space in order to be engaged in the change
• Lack of support from employees	

Why Organizations Fail at Change

Organizations frequently fail at change for a number of reasons, as shown in Table 2.8.

Consider again the All Company Training example. In this example, All Company Training leadership is mitigating potential failure by engaging employees early on and asking for their input pertaining to the decision of what products to produce. This increases the support from employees and therefore the likelihood that they will participate actively in the initiative throughout its implementation. Consider this example of a client that failed on a major strategic change initiative.

> A U.S.-based retail organization needed to make major changes to business processes. This need came about due to substantial customer losses because of significantly outdated processes. The customer losses had drastically impacted revenue over the last five years—a loss of over $1 billion. Unfortunately, prior to addressing this strategic need, the organization laid off one-third of their workforce, the majority of whom worked with the customers. This major transformational change initiative, therefore, was launched with employees who had limited knowledge of the customers and were angry about the layoffs. They were not only taking on more work to make up for those employees who had been laid off, but were also now being asked to work on a change initiative with a very short timeline. Additionally, many of the remaining employees were upset because they perceived that had the organization reacted sooner to the situation, their colleagues might still have jobs today.

In this example, this organization was, frankly, destined to fail. The organizational leaders did not have the support of their employees; they were reactive to a situation that they could have been proactive about, had they paid attention; and they were not considering that other happenings in the organization (e.g., workloads being shifted to make up for a reduction in staff) would impact the ability of employees to commit to this initiative.

Organizations will fail at change when they haven't created a desire for change within the organization. This desire can be created when the organizational leadership invests time in ensuring understanding of the change and sharing a vision for that change. Additionally, ensuring involvement in shaping the change by those who will be most impacted by it and will have to work with the change is essential to creating that desire to change.

One project manager shared this unfortunate experience. He was assigned to lead an organizational change project that initiated discovery on reasons for employee disengagement through an anonymous survey. The project sponsor—a senior leader in the organization—then opted not to share the results of the survey, nor acknowledge the information gathered, but instead, completely disregarded it while executing the change initiative. As you would guess, lack of acceptance by employees from throughout the organization was the result of that change project!

The impacts of change are different from one part of an organization to the next. If leadership does not recognize these impacts, the change initiative is destined to fail. Not addressing an impact in one department means there will be trickle effect throughout the organization. Even if a particular change initiative will only be focused on one department, it is essential to consider the organization *as a whole*. Departments do not work in a silo and not recognizing that change in one department will have a trickle effect in another area of the business will only increase the challenge of trying to accomplish that change.

Allen, the SVP of Finance, wanted to make changes within the finance group. As the department had grown over the years, inefficiencies in getting invoices paid as well as in collecting monies from customers had become even more impactful to the bottom line. Additionally, rather than hiring full-time employees, the group was increasingly relying on contractors to assist during certain times of the year. Allen wanted to restructure the group to increase efficiencies and reduce the reliance on contractors within the group. He proceeded with the design of his restructure of the organization and presented his plan at the next senior leadership meeting. He was met with significant resistance. Three other department heads reminded him that his changes—while likely beneficial to his department— would impact how their departments interacted with Finance. Thus, it was back to the drawing board for Allen.

Consider the situation with Allen. Allen forgot two important considerations for change: (1) his finance group does *not* exist in a silo and (2) the success of his reorganization *requires* the support of his colleagues. Allen may have created the best organizational structure, but without sharing his ideas and accomplishing his desired change in collaboration with other departments in the organization, Allen's restructuring cannot succeed.

Failure is also seen in change initiatives when the change is too big and leaders, as well as employees, just can't see how to approach the change. When a change is very complex and large, it is difficult to see progress quickly enough to keep employees engaged in the change. Without looking at all that may need to be accomplished to succeed with the overall desired change, the organization risks failure of the change. Let's look at another example.

> An organization wants to speed up the decision making within the organization. This is a major change within the organization as, currently, decisions are made only by senior leadership. It will require many other changes before decision making can be pushed down to the lower levels within the organization. The organization will also need to undertake change initiatives to accomplish the following:
> * Restructure the organization to enable decision making at lower levels
> * Establish roles and responsibilities around decision making
> * Determine current skills of employees as they relate to making decisions
> * Train employees in best practices for making decisions and a methodology for decision making
> * Create new processes and procedures in order to ensure reduced risks in decisions made, as well as consistency in decision making
>
> The organizational leadership will need to determine the priority of these various change initiatives in order to ensure that the organization can eventually accomplish the goal of pushing decision making down in the organization and thereby speeding up decision making.

EFFECTS OF CHANGE ON THE INDIVIDUAL EMPLOYEES

Organizations cannot solely be concerned with the effects of change on the organization; they must also be concerned about the effects of change on individual employees. Fear and frustration around change is natural in every human being. Even though some employees will say they are not worried about change—assume they are worried. Organizations tend to forget that change affects individuals in a variety of ways and that perceptions of change must be managed *prior* to the change happening. This tendency is more common when a change is perceived by the organizational

Table 2.9   Effects of change initiatives on individuals

• How the individual performs the job, processes, and procedures to follow • The specific job, role, or responsibilities the individual will perform based on the change • Current skills or knowledge of the individual	• Whether or not the individual will still have a role in the organization once the change is implemented • Hours the individual may need to work to fulfill his/her responsibilities • The motivation of the individual

leadership as insignificant or having a minor impact. What one individual considers insignificant or minor, another may find overwhelming. While organizational leadership cannot completely eliminate frustration and fear around a change that will happen, they can manage that frustration and fear. This is done through building trust between employees and organizational leadership by sharing information, the vision for the change, and by ensuring sufficient and effective communications. Table 2.9 provides some effects of change on individual employees.

As can be seen in Table 2.9, it is essential to understand how individuals within the organization will be impacted by the change initiative in order to engage them in that change. Even though a change *appears* to be minor, each employee will react differently to that change depending on a number of personal factors, such as:

- New skills that may be required for the individual to be successful in the role
- The fear of the unknown—what specifically does the change mean for the individual
- The individual employee's personal history around change initiatives within and external to the organization
- Whether the organization and its leadership has been perceived to be truthful and/or has provided all of the information available about the change
- What else is going on in the individual employee's personal life (*outside of* the workplace)

An employee, for example, who is older and has been working in the same role for 20 years or more may focus on new skills they need to learn to work with an implemented change. Even if the organization will provide training for the employee, there is a fear of learning a new skill and, in particular, whether or not the individual will be capable of learning the skill. Certainly not every employee in the same role for 20 years will feel exactly the same way—but understanding these perceptions and these fears around change will lead to more effectively addressing the workforce overall and ensuring that they adopt the change.

> *A global firm merged divisions approximately six months after an acquisition. I worked with the new leadership team to facilitate a conversation around the vision for the division and to craft communications to share that vision with all impacted employees, as well as the organization as a whole. The communications initially were focused on sharing information about the merger—specifically, who, why, what, and when.*

> *Of particular concern to leadership was the fact that there would be layoffs due to this merger and there needed to be transfer of work as well as cross-training prior to layoffs occurring. Given that layoffs were expected by employees, it did not make sense to hide the fact that layoffs had to happen. Early communications acknowledged that there would be layoffs and shared with all employees what the organization was doing to support those being laid off. This allowed the organizational leaders to move past the focus on who was losing their jobs and look forward to the opportunities the merger opened up for the organization.*

By acknowledging that there would be impact from the change, and that it would be a negative impact (employees would lose their jobs), the organization was able to keep the workforce engaged and moving forward throughout a significant change initiative. Often, organizational leadership attempts to hide the fact that layoffs may occur due to a transformational change initiative, but this is never successful. This makes the assumption that individual employees do not know what is going on and are not intelligent enough to figure it out.

OBSTACLES TO CHANGE

Obstacles to change occur at both the organizational as well as the individual level. Even if change is *good* for the organization, there will be obstacles to overcome on the way to achieving that change. This is thanks to human nature and the desire to keep the status quo. Organizations will have a difficult time reaching a successful conclusion to their change initiative if they assume that because a change is good for the organization, it will be seen as a positive initiative overall.

A colleague of mine was leading a business transformation project that was focused on transforming how a bank serviced its customers via several delivery channels, including internet banking and mobile banking. The project was launched to establish a brand new customer contact center, which resulted in the introduction of new business processes, service levels, problem escalation, and resolutions. Challenges for the project team were centered on managing the effect of new business processes and changes in organizational structures, as well as sourcing individuals who fit the corporate culture to fill the role of customer contact personnel. This individual shared that the team overcame these challenges through continuous training and development and through the engagement of contractors to secure the best candidates who fit the culture of the bank. Executive management also remained engaged throughout the initiative, ensuring alignment with organizational strategy and removing barriers.

Table 2.10 provides a list of some potential and common obstacles to change initiatives by employees within the organization.

The obstacles shown in Table 2.10 are aligned with the impacts of change on the individual employee (Table 2.9)—there is overlap in both tables. When organizational leadership considers and addresses obstacles toward implementing change, they will also address those areas of impact of most concern to the individual. Obstacles and impact go hand-in-hand.

A manufacturing organization needed to make significant changes in their processes in order to reduce costs and improve time-to-market for their products. Redundancies in processes that have occurred over the 20 years that the organization has been in business were

Table 2.10 Common obstacles to change

• Fear of the impact of the change on the employee and his/her role and responsibilities	• Disruption in the employee's daily routine
• Fear of losing control of one's work and role	• A "we have always done it this way" mentality
• Insufficient information about the change and why it needs to occur	• Viewing the change as inconvenient given all else going on in the workplace or the employee's personal life
• Fear that the change will impact the employee financially	• Fear of job loss
• Change fatigue	• Tools and technology insufficient to implement the change

having a significant impact on the bottom line. The organization saw the benefit of making changes in processes in order to increase revenues and profits. The assumption was that employees would understand that increased revenues and profits would benefit them also. The communication to employees, via e-mail by the head of manufacturing, stated, "Because of inefficiencies in how the work is done, the organization is being impacted negatively. Changes must happen in manufacturing processes, and they are a priority for the organization."

In the manufacturing organization example, the head of manufacturing has not sufficiently communicated with the employees to enable them to embrace the change. The focus was on the organization only and seemed to imply that the inefficiencies were the fault of those working with the processes. This message alone signifies the following potential effects of change on the employees (see Table 2.9):

- How the employees perform their job
- The specific responsibilities of the employees
- Whether or not employees will still have a role in the organization
- The current skills and knowledge of the employees

Additionally, this statement by the head of manufacturing introduces, or feeds, the following obstacles to getting employees to come along with the change (see Table 2.10):

- Fear of the impact of the change
- Fear of losing control
- Fear that the change will have a financial impact
- Disruption in the employee's daily routine
- Fear of job loss

At the organizational level, obstacles to change are often around a lack of savvy on the part of leaders as to how to effectively lead a change initiative. This includes an arrogance in assuming that leaders alone can push forward change based on the fact that change is good for and benefits the organization. This is a fallacy. Unless and until the organization can get the employees—those individuals who actually *do the work* of the business—to understand and embrace the change, the organization cannot possibly be successful in the change. Smart executives have a project manager in place to lead strategic change initiatives with the support of top leadership.

Ideally, this individual has strong relationships throughout the organization in order to engage all employees in change.

> *Another colleague of mine shared that he is often brought into an organization to manage projects that are failing. He often found that failure was due to a few common issues, such as:*
> - *Poor management at the leadership level*
> - *Lack of communication*
> - *Fear of employees to raise issues or problems*
>
> *He worked with these organizations to change how such projects were managed by ensuring a strategic project management approach and strong communication channels throughout the organization.*

Many experts consider change fatigue to be one of the biggest obstacles to change. This often occurs when organizations run change initiatives back-to-back or when change is not well-thought-out and initiated only to be abandoned and attempted again. A survey of 2,200 employees (including executives) by the Katzenbach Center noted that change fatigue is a major obstacle to successful change (65% of respondents commented that they felt some level of change fatigue). Change fatigue, while a common problem can be resolved when organizations:

- Align changes to strategic goals
- Implement changes in a controlled, process-driven way
- Utilize strategic project management to manage the change initiative from start to finish
- Plan up front for change, considering the impacts of the change on the individuals and all else going on within the organization

OVERCOMING OBSTACLES TO CHANGE

Obstacles to change are best overcome and managed through communication. Communication that occurs early on (prior to the actual launch of the change initiative), continues throughout the implementation of the change, and continues *after* implementation of the change is the best way to overcome many of the obstacles shared in Table 2.10.

Table 2.11 provides a variety of ways to overcome obstacles to change at the organizational, business unit/divisional, departmental, as well as the individual levels.

All of these methods to overcome obstacles to change are focused on effective and sufficient communications throughout the organization.

Table 2.11   Ways to overcome obstacles to change at various levels

At this level...	Do this...
Organizational/Business Unit	• Highlight what happens if the organization does not change (organizational/business unit impact) • Clarify the change initiative to ensure understanding • Ensure a vision that is clear, exciting, and achievable
Departmental/Workgroup	• Highlight the improved support of clients through changes within departments or workgroups • Highlight how projects will be better accomplished by workgroups • Highlight departments or workgroups already working within change or a component of change and their success
Individual	• Highlight the long-term impact to individual employees if change does not happen within the organization (if the organization closes, employees lose their jobs) • Share benefits of change in helping to make the job easier/make the work better • Promote the ability to learn new skills and increase knowledge • Promote the excitement of something new • Ensure individuals are involved in defining the change and/or implementing the change

As a best practice, when working with clients to engage their employees in change, I prefer to start conversations with employee groups by asking them to share:

- *What they believe may or may not work with the change initiative*
- *How they would proceed with the change if they were the ones in charge*
- *What they would want to know, or need to know, in order to alleviate their concerns*
- *What success looks like to them*

For individual employees who continue to resist change and don't want to participate in sharing of information, I converse with them one-on-one to understand more about their concerns that are causing resistance.

Stakeholder Support Committees are one of the best groups to assist in removing obstacles to change. First, these committees are comprised of individuals from throughout the organization who are impacted by the change and may have to overcome their own personal obstacles to change. They carry significant weight with other employees because they have already built trust and have established relationships with their colleagues and coworkers.

> *I utilize Stakeholder Support Committees for all of the transformational change initiatives that I am leading in order to engage employees throughout the organization. These committees are comprised of other employees from all levels—individual contributors up to supervisors and mid-level managers. They are tasked with sharing information about the change initiative, answering questions, and bringing concerns back to leadership to be addressed. Because they are considered "one of them," employees throughout the organization are more likely to share concerns and listen to the committee members over senior leadership.*

More information around Stakeholder Support Committees and their value in transformational change initiatives will be shared in Chapter 3.

A FRAMEWORK FOR LEADING ORGANIZATIONAL CHANGE

Recall that Figure 1.2 depicted a high-level framework for leading organizational change. Leaders who understand change, its impact on the organization, as well as, and perhaps even more importantly, on the employees, often have key steps they take for leading change in their organization. These steps are consistent whether it is a simple change or a large-scale change effort. The amount of effort, however, applied to each step may vary based on the complexity. Figure 2.7 provides more details of each of the steps first outlined in Figure 1.2.

For example, for a smaller, less complex initiative that may only impact a couple of departments or workgroups in an organization, there is no need for a Stakeholder Support Committee. The function of this group may be replaced with a regular conversation about the change led by department heads. There still needs to be communication, but given the limited individuals impacted, the communications are more easily managed without the need for a Stakeholder Support Committee.

Figure 2.8 provides a high-level timeline for a preplanning effort by an organization for a strategic, complex change initiative.

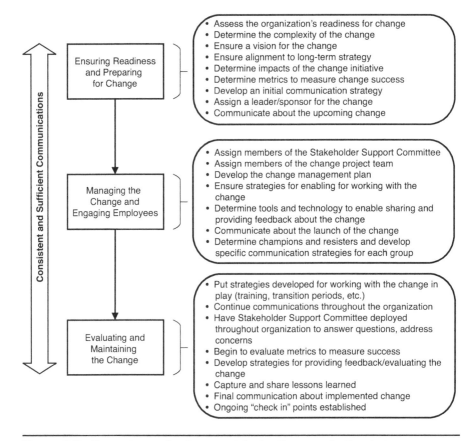

Figure 2.7   Expansion on the framework for leading organizational change

As can be seen in Figure 2.8, this organization began planning in January for a complex, strategic change initiative that was kicking off in September. Early planning was heavily focused on communicating about the initiative and beginning to get buy-in from employees. This required the organization to share the vision for the change and consider the impact to both the organization as well as individual employees. This information helped to craft early communications to engage employees. March through April was reserved for socializing the change initiatives throughout the organization via a number of Road Shows. These were led by executives and other senior leaders with the intent to engage employees in the effort, ensure understanding of the change, as well as determine employee concerns that needed to be addressed to ensure a successful change initiative.

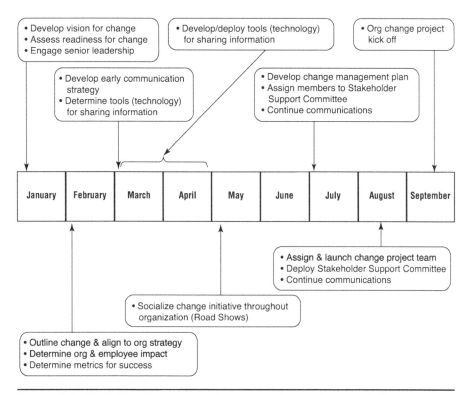

Figure 2.8   Partial timeline: preplanning for organizational change initiative

This book has free material available for download from the
Web Added Value™ resource center at *www.jrosspub.com*

3

LOOKING AT CHANGE FROM A POSITIVE PERSPECTIVE

"Learn to adapt. Things change, circumstances change.
Adjust yourself and your efforts as to what is presented
to you so you can respond accordingly. Never see change
as a threat, because it can be an opportunity to learn,
to grow, to evolve and become a better person."
Rodolfo Costa, *Advice My Parents Gave Me and
Other Lessons I Learned from My Mistakes*

In an ideal world, every leader and employee would look at a change from a positive perspective. Change would excite us! We would see the benefit and value of the change. We would be thrilled to be involved in making something better or creating something new. We would—regardless of our role and our responsibilities—initiate and drive change to improve the way we work and the way the business runs. We would see change as a regular occurrence rather than a chaotic situation. But that is in a perfect world and, sorry, that doesn't exist. As human beings we often resist change; we rebel against it. We prefer the status quo—it's easy, and we are used to it. Changing requires changing our behavior. That is difficult to do, and we are hesitant to do it.

*On a biannual basis, during all-staff meetings, I work with one
of my clients to engage employees in change through asking them*

questions to get them thinking about the organization and how to improve the procedures used to perform their tasks. For example, at the last all-staff meeting, the conversation started a week before the meeting with the following question posed on their internal website: "We know there are some changes coming our way. If we moved toward them rather than letting them come to us, how would that look?" By the day of the all-staff meeting we had already received over 100 responses from employees throughout the organization. The responses provided leadership with some great insight into the employees' perspective of what was happening with customers, products, as well as vendors and suppliers; including thoughts on how to fix some issues. Employees were, effectively, proposing the change for the organization! This activity has been ongoing within the organization for the last two years and has, over the years, made change a more positive experience for individuals since they are the ones shaping the change.

Sponsorship for the change at the leadership level will help to build commitment for the change throughout the organization. Executives *must* be actively involved in framing that change and in driving that change throughout the organization, if it is going to be successful overall. While a smaller change initiative—such as one that impacts only one department or workgroup—may be easily sponsored by a department head, a large, complex change initiative that will have impact throughout the organization must be sponsored by senior executives within the organization.

This chapter will focus on how leadership can assist in creating an environment for looking at change from a positive perspective through communication to engage and excite employees around the change.

NEGATIVE CHANGE PERSPECTIVE VERSUS POSITIVE CHANGE PERSPECTIVE

Negative change perspectives are common and should be considered to be prevalent when change is first initiated within an organization. This will be especially true if any of the following is considered the norm:

- There is a culture of the status quo within the organization
- There is a history of unsuccessful, unsupported change initiatives
- There is a history of change initiatives being launched, canceled, relaunched, and canceled again
- There is limited availability of resources to support the change or change is launched in the middle of a number of key projects in progress

- There is limited knowledge of how to lead change at the executive level
- Change initiatives do not follow any structured approach and are haphazard and launched without consideration of what else is going on in the organization
- There is a history of poor communications within the organization or a perception of leaders being dishonest

These negative perspectives are indicative of any or all of the following employees' reactions:

- A focus on everything that could go wrong with the initiative
- Insisting that all is fine as it is and it doesn't make sense to change
- Being disruptive during times of change; gossiping about the change
- Highlighting any number of problems with the change but not providing solutions
- Actively working against the change initiative
- Refusing to participate on change teams

While in the previous list the focus is on employees, keep in mind that this same negative perspective could reside with leadership. Leadership is not immune to feeling negative about change. They are human beings after all and have all the same worries and concerns and fears about change and how it may impact them.

> One Chief Executive Officer (CEO) who was excited about launching a change initiative that would increase innovation in his organization was shocked that his senior leadership team—which included the Chief Financial Officer, the Chief Technology Officer, and the Chief Marketing Officer—were against the change. They were not open about their resistance when he shared the change and its benefits with them, but rather he learned about it two months after launch that they were actively working against it!

A key need to be able to adapt to change is to be resilient. Individuals who are resilient are more able to adapt to change. *Resiliency* is defined by The American Heritage Dictionary of the English Language as *the ability to recover quickly from… change…; buoyancy*. Resiliency provides individuals the ability to bounce back or to recover from change. Individuals who are considered resilient are more likely to see the positive side of change and stay focused on the possibilities and opportunities rather than to get absorbed in negativity.

Resilient employees also have less stress overall. Since change creates stressful situations, being resilient is a benefit for employees. These are individuals who do not shy away from change. Rather they view change as an opportunity. When organizations are managed by leaders who are resilient, employees are more likely to adapt to change more easily.

> *One of my clients has leadership in place that embraces change and makes it exciting for employees through regular conversations around change. Additionally, they prepare their employees for change through providing regular opportunities to develop skills and build knowledge via training programs, attendance at industry events, and by providing mentors. This enables their employees to develop those future skills and knowledge they will need in order to adapt quickly to change.*

Leaders who are resilient, and therefore adapt to change more easily, are better able to engage employees in change. Those leaders will help employees to adapt to change when they:

- Share opportunities for change and ask for input/feedback from employees
- Identify where threats to change might exist and determine strategies to address those threats
- Understand that there will be resistance to change and work to address that resistance
- Never assume that change will be easy, even if it is a *simple* change
- Keep employees informed about the change and progress toward implementation

Leaders who understand the need to encourage and support resiliency around change realize that creating a climate in the organization that supports and enables change is essential to developing that resiliency among their employees. This requires ensuring a thorough understanding of the change and sharing the vision for the change. A focus on the *why* and the *benefit* of the change will assist in promoting resiliency in change. Communication—early on, throughout, and after implementation—is essential to promote a comfort level and therefore resiliency in change.

> *Jeremy noted that Leila, one of his top employees, was also engaged in any change that was launched within the organization. Jeremy could count on her to take the lead in promoting the change to her coworkers. She looked at change as an opportunity to do something*

different—to improve, to work more effectively, to solve a problem, or realize an opportunity. Whenever a change was launched, Leila seemed to have a natural ability to link that change to how it would solve a problem the team was having, enable them to realize an opportunity, or empower them to better support clients with less effort on their part. Her excitement about change seemed to be contagious among her coworkers.

Negative perceptions also occur when employees' frustrations are allowed to grow due to inaction, and the perception of unconcern, by leadership. Consider this story:

A friend and colleague of mine worked in the U.S. for a European company that has satellite offices in over twenty countries. This company is over a century old and, unfortunately, shows its age in the implementation of processes. The U.S. team was tasked with delivering a multimillion-dollar project portfolio to several cities across the country. There was a very well-defined set of processes in place; there is no doubt—a process for this, a process for that, bridging the business needs across the Atlantic in the manner it could. The large majority of customers were highly dissatisfied with the response and service they were getting. Employees in the U.S. were quite frustrated with their inability to effectively do their job—painfully manual processes and institutionalized silos were the norm. Finally, the logical outcome of these issues resulted in contractual complications that could easily have been avoided. The U.S. team attempted to make changes, but the European headquarters would not allow it. Where this story gets interesting is that headquarters was not averse to changing, they were just not aware of the conditions on the ground in several countries. Manual, opaque, and over-designed business processes that introduced many unnecessary steps were standing in the way of the satellite offices being able to serve their customers. The company said the customer was number one but the processes told a very different story. The moral of this story is that it is very important for a company, especially a multinational, to always remain open to soliciting honest feedback from the regional offices in order to understand what their employees are experiencing on the ground. This company was in the process of losing their good people who had grown weary of banging their heads against the wall.

As first shared in Chapter 1, change is an emotional process. Every employee deals with change differently. Some will accept change more readily because they see change from a more positive perspective; other employees drag their heels when change is happening because they see change as a negative. If change is launched with the fact that employees will have to be engaged in understanding and accepting that change as a priority, organizations are more likely to spend the time necessary to engage *all* of the employees in the change initiative.

Figure 3.1 provides one potential emotional process through which employees may move as they first learn about change through to understanding more about the why of the change and *seeing* the vision for change.

Let's look at Figure 3.1 in more detail using a case study example:

All Company Training is launching a new change initiative that will impact their Operations Group. An employee in the Operations Group hears about the change that will be launching from his manager. His manager seems excited about the upcoming change. The employee initially feels positive about the change (anticipation of something good in the future). Based on what he has heard from his manager, it seems that the change will enable work to be accomplished more easily and he may get to learn some new skills, too. About three weeks have passed since the employee has heard from his manager and the change is already in progress. The employee hasn't been asked to participate and still doesn't know if the company will train him to obtain the new skills he'll need to be successful. He is beginning to wonder if maybe he is going to lose his job. The employee has a negative perception of the change and is feeling apprehensive. Finally, the employee decides to ask what is going on. He talks with his manager and learns that while the change initiative is in progress, it is in early stages. The manager tells him more about the change and why it is happening. It is obvious the change

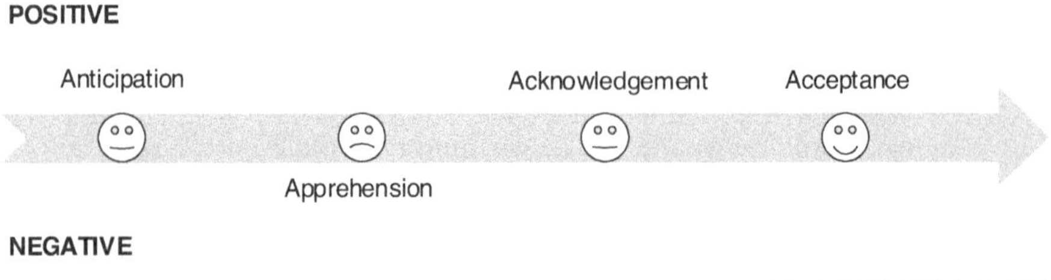

POSITIVE

Anticipation   Acknowledgement   Acceptance

Apprehension

NEGATIVE

Figure 3.1   The positive and negative perceptions of change

will not only be beneficial for the organization but will also be beneficial for employees. The employee also learns the following:

- *A meeting will be scheduled in another week to get input from employees*
- *Training programs are being developed by Human Resources to ensure that employees will have the skills needed*
- *There will be a transition period so that employees have time to ensure they know how to work with the change*

The employee has moved to acknowledgment of the change. He is feeling more positive about the change and looks forward to the upcoming meeting. The meeting occurs, and the employee learns even more about the change initiative. He has been asked to participate in the change initiative by sharing his feedback and by being a part of a pilot group. The employee is feeling very positive about the change initiative and is in acceptance of the change.

As mentioned earlier, this is one potential emotional path to accepting change for employees. Another employee may start off his emotional path with negative feelings about the change and may retain that negative perception for a longer period of time. Some employees may move to *acceptance* only after the change has been implemented, and it is going well.

Sharing the positive aspects of change with employees will allow you to swing their perception from a negative to a positive. In the case study example previously mentioned, the manager shared more information that increased the comfort level of the employee and persuaded him to feel more positive about the change. It is possible that, had the manager shared information sooner—before being asked by the employee—the employee would have moved from *anticipation* to *acknowledgment* and skipped being *apprehensive* about the change.

Positive aspects about change that may be shared to encourage an improved perception of change include:

- New and better ways to get work done
- New opportunities for employees (growth potential)
- Improved ability to satisfy the customer
- Development of new skills and knowledge (personal development)
- The ability to be involved with new or cutting edge technology
- The ability to be involved in shaping the change

Organizations often see a more positive perspective about change overall when employees:

- Are engaged in the change—excited about the opportunity and providing significant feedback (involved in shaping the change)
- Are willing to let go of the past and focus on the future (move past *that is the way we have always done it*)
- Ask questions early on about the change as they attempt to understand it and get honest and straightforward answers from leadership
- Contribute to change within the organization by regularly striving for better ways to get work done

The Value of a Positive, Opportunity-Driven Perspective

Looking at change from a positive, opportunity-driven perspective is of value to both the organization and the employees. Figure 3.2 provides a checklist of items that indicate a positive, opportunity-driven perspective of change within the organization.

These statements in Figure 3.2 are all aligned with organizations and individuals who view change as positive and full of opportunity. An organization that cannot check off each item on this list as how things work within the organization, would want to develop a strategy to improve how change is viewed overall—from the senior leadership down to each individual contributor. In nearly all organizations, if senior leadership supports

Items indicative of a positive, opportunity-driven perspective of change…
Change is seen as an opportunity for personal and professional growth, enabling for new skills and building knowledge.
Change means that the organization and its employees are moving forward and accomplishing their strategy.
When change is happening, employees ask questions, provide feedback, and expect to be involved in the change initiative.
Employees take the lead to initiate change, frequently making positive changes in how they get work done.
Leaders share opportunities for change with employees as well as share how change can help address threats to the organization.
Leaders regularly reach out to employees to ask them what the organization can do to better support them and customers.
Leaders regularly engage employees in change through asking thought-provoking questions.
The face of change initiatives within the organization are the employees, not the leaders.

Figure 3.2 Checklist indicating positive, opportunity-driven perspective of change

change and encourages participation in change, employees are likely to follow suit.

> *Sally was the VP of Marketing in a large retail organization for the last 10 years. In all of that time, she has never initiated nor supported change within the group. Even the simplest request by Sally's manager to change a process was met with resistance by Sally. Her employees couldn't even suggest changes to processes. While her manager could understand that people are not always comfortable with change, it seemed that Sally just plain hated any kind of change and always had a negative perspective. It was beginning to trickle down to her staff.*

In this scenario, Sally consistently views change as a negative; something to be avoided at all costs. This view of change has a major effect not just on Sally's professional development and potential within the organization, but also on the engagement of her staff as well as the ability of the organization to move forward.

> *Sally was now faced with reporting up to new senior leadership. In particular, the new Chief Marketing Officer for the organization was a big proponent of change and enabling employees to participate in that change. In fact, everyone on the new senior leadership team had a vision for change. At an off-site meeting of all leadership, including Sally and her peers, the need for change aligned to a refocused strategy for the organization was discussed. Numerous times during the meeting Sally said that things were fine as they were and that Marketing certainly did not have any need to change. She also commented that she had no time to work on change initiatives, nor did her team—they were simply too busy. Three months after new senior leadership took over the organization, Sally lost her job. The reason: she was holding back the Marketing Department and her staff by refusing to move forward and change. The Chief Marketing Officer told Sally that change was necessary for continued growth and he needed someone leading the team who understood the need for change and who would embrace change.*

Drivers and Resisters to Change

Not every individual in the organization—including leadership—will support change from the start. It should be expected that resistance exists within the organization and that it must be addressed and managed. Certainly we must initially get support at the top of the organization because without that support, it will be difficult, if not impossible, to engage the lower levels in change. Table 3.1 depicts a number of drivers and resisters to change.

When an organization recognizes what will cause resistance to change, they can ensure that those issues are addressed *prior* to initiating the change project.

> *Alexis and his team were excited about the change that was launched by the VP of Technology. This change generated a number of benefits for their group, including:*
>
> * *Training to gain new skills in a cutting-edge technology*
> * *Opportunities to be involved in using that new technology to launch a new product that would be the first of its kind in the marketplace*
>
> *Additionally, for Alexis, there would be a promotion involved, which would enable his team to take on additional responsibilities—effectively enabling growth opportunities for everyone.*

A best practice for any leader who is launching change is to put themselves into the shoes of their employees *prior* to actually launching the change. Leaders may list the drivers that will enable employees to embrace the change and what will cause employees to resist the change. If this

Table 3.1   Drivers and resisters to change

Drivers that encourage employees to embrace change...	Resisters that encourage people to fight against change...
• Personal and/or professional benefits and opportunities are tied to the change • An understanding of the need to make the change • The change enables challenges in work assignments that encourage individual learning opportunities and new skill development • There is more autonomy granted to the individual in his/her role as a result of the change; the individual's expertise is valued	• No personal or professional benefits and no opportunities as a result of the change • No clear understanding of why the change needs to happen • The change does not allow for challenging work and, in fact, appears to make the job boring • The individual loses status in the organization; his/her expertise is not valued and there is less autonomy in the role

information is considered and understood up front—especially at a high level—it is much easier to craft communications and other ways to engage employees and address that resistance while highlighting the benefits.

> *A global organization merged divisions approximately six months after an acquisition. I worked with the new leadership team to facilitate a conversation around the vision for the new division and to craft communications to share that vision with all impacted employees and the organization as a whole. The communications initially were focused on sharing information about the merger—specifically, who, why, what, and when. Of particular concern to leadership was the fact that there would be layoffs due to this merger and there needed to be transfer of work and cross-training prior to layoffs occurring. Given that layoffs were expected by employees, it did not make sense to hide the fact that layoffs had to happen. Early communications acknowledged that there would be layoffs and shared with all employees what the organization was doing to support those being laid off. This enabled them to move past the focus on who was losing their jobs and encouraged looking forward to the opportunities the merger opened up for the organization.*

By acknowledging what was already common knowledge, rather than trying to hide it (as too often is the case), the organization addressed a large point of resistance to the change; enabling them to move forward. Had the organization chosen to ignore the fact that there would be layoffs, employees would have been focused on the layoffs, gossiping about what was going to happen, which would have had a negative trickle affect throughout the organization. If leaders do not address resistance when it arises, they only create more defiance over the change, with individual employees actively working to sabotage the change initiative.

The most effective way to manage through resistance and encourage drivers so employees can embrace change is through regular and consistent communication. Resistance can also be addressed when leadership:

- Ensures involvement by employees in the change initiative through asking for feedback
- Ensures that plans are in place to enable the employees to be trained to learn new skills
- Understands what is concerning to employees about the change and addresses those concerns

It is essential to address fears to gain support and commitment for the change initiative. If fears are not addressed, the organization is unlikely to have a successful change initiative. Even if in the end it appears to be successful, the organization has likely created significant mistrust between leadership and employees, which creates a disengaged workforce.

COMMUNICATION BEST PRACTICES

Earlier chapters have touched briefly on communications. Let's explore this topic more fully in this chapter. Communication is the core of any successful organizational change initiative. It is, without a doubt, the foundation to build and encourage positive organizational change. Communication enables the organizational leaders to frame the change in a way that helps employees from throughout the organization understand the importance of the change initiative and how they can help in achieving success. Effective communication—right from the very beginning—enables leaders to influence how the organization will view the change.

> *One leader of a global organization brought the members of the Project Management Office together at the start of each fiscal year to share with them the types of change initiatives that would be launched for the upcoming year and why they were necessary for the success of the organization and for meeting long-term strategic goals. This leader knew that he needed to engage those who would be leading these change projects in order to ensure that they were interested in and understood the reason for the change projects. In this way, they would support the change projects with their teams and throughout the organization.*

The willingness and ability to communicate constantly and honestly about change is essential to the success of change initiatives. Communicating early on about a change initiative makes it possible to identify negative impacts of the change and allows for addressing those negative impacts to ensure overall change success. Figure 3.3 provides a checklist of items to consider for inclusion in initial communications about change.

The more comprehensive the initial communication, the more likely that employees will feel comfortable with the change, even if they have further questions. Figure 3.4 is an example of a partial initial communication inviting individuals to a meeting to learn more about an upcoming change.

	Can the following be included in early communications about an upcoming change initiative...
	A vision for the change
	The expected impact of the change – positive and potential negatives and how they will be addressed
	A plan for future communications and channels for communications
	Feedback mechanisms for employees to provide ideas, opinions, thoughts, suggestions
	Stakeholder Support Committee members
	An initial plan to launch the change (broken down into smaller components if complex)
	Information about a forum for Q&A after the initial communication about change

Figure 3.3   A checklist of items to include in early communications

> All Company Training is pleased to announce the long-awaited launch of the organizational change initiative!
>
> Join the CEO in the Cafeteria at 8:00 AM on Wednesday, Noon on Thursday, or at 4:00 PM on Friday to learn about the benefits of this initiative.
>
> Bring your questions, concerns, ideas, and suggestions. Interested in learning more before then?
>
> Drop by the CEO's Office anytime on Monday or Tuesday!

Figure 3.4   Initial communication to learn more

Let's explore a bit more of the background information on Figure 3.4. The CEO knew that there would be concerns about the change initiative. He had heard from many of his senior leaders that employees were wondering about the impact on themselves and the customers. He wanted to provide a forum where they could get questions answered and he could address concerns. He provided a number of options—morning (and served breakfast), noon (and served lunch), and then later in an afternoon (where he served appetizers and had beer and wine available). He also knew that some employees would be unlikely to speak up if there were many individuals around, so he invited those employees to drop by his office. The follow up to this initial communication about the change was small sessions led by the CEO as well as his senior leadership team, where employees were able to engage on a smaller scale and continue to learn more and get answers to any additional questions or concerns.

The Five Ws—*Who, What, Why, When,* and *Where*—is a great tool to use when sharing information about change to ensure that the key points about the change initiative are covered in any communications or conversations. The Five Ws is a common tool used frequently in any number of situations. As it relates to change initiatives, it ensures that the organizational leadership is answering the key questions that will arise from throughout the organization about change. Figure 3.5 provides a checklist to make sure that leadership can respond to key questions regarding change *before* launching and communicating about that initiative.

When leadership takes the time to ensure that there is a response to each of the Five Ws as shown in Figure 3.5, they are well on their way to answering the *most pressing* questions that will be asked by employees about the change.

Since communication is absolutely key to successful change initiatives, the ability to answer questions early on is necessary to engage employees. If that initial communication does not address key *who, what, why, when,* and *where* questions that are in the heads—if not out of the mouths—of all employees, the change effort will be off to a difficult start.

A national retail organization's change initiative for the upcoming year was to implement the use of technology for salespeople out in the field. Leadership knew that the salespeople were hesitant about the use of the technology because they felt it was being used to track their movements when out of the office. The head of sales prepared

Can you answer the following questions about change...	
	WHO will be impacted by the change?
	WHAT is the change?
	WHY does the change have to happen?
	WHEN will the change happen?
	WHERE will the change specifically take place?

Figure 3.5   A checklist to ensure key questions about change can be answered by leadership

for the initial conversation about the upcoming change initiative by responding to the following questions:

- *Who: The change is impacting every sales associate in the organization.*
- *What: The change is the implementation of technology for use by sales to enable more effective tracking of individual clients as well as having access at the fingertips of sales to information about clients—such as spend, quantity of product, types of products, etc.*
- *Why: As the organization continues to grow and expand and new sales associates come on board, it is essential to facilitate better sharing of information and tracking of customers. It is more difficult today than it was just a year ago to understand the buying potential of all of our customers and to ensure that we capitalize on that potential. The benefits are great for sales associates, as they will have all the key information about their accounts right at their fingertips, wherever they might be.*
- *When: The change initiative is kicking off early in the year during the all-sales meeting to be held in Orlando, FL. At this time, sales leadership will explain more, as well as ensure sufficient time for discussions around the impact of this change. Prior to the kick off, sales leadership will engage a project manager to help in ensuring the change project is implemented using project management best practices.*
- *Where: The change will take place throughout all of our sales teams in all locations across the U.S.*

In this scenario of the national retail organization's sales team, the head of sales answered the Five Ws (key questions about the change) that will be used in her initial communication with all sales employees. While certainly the answers to these five questions may not address every single question that sales employees may have about the initiative, they cover the most pertinent questions that are common at the start of every change initiative. Having the answers to those questions ready—whether the communication is via e-mail, during an all-staff meeting, or via another forum—will tend to increase the comfort level of those impacted by the change because there is a perception that the change has been considered carefully at the leadership level.

Developing the Communication Strategy

A key document in any change initiative is the strategy behind how communication will happen within the organization. Figure 3.6 provides an example of a partially completed template for a communication strategy for a multidepartment impact process improvement change initiative. This same template is available as a downloadable file from the Web Added Value™ Resource Center at www.jrosspub.com/wav.

Many organizations, especially larger ones, have standards around how communication occurs within the organization, as well as responsibility for communications. Figure 3.6 is one example of a partially completed communication strategy for a client initiative. The project was scheduled to kick off in January; however, as can be seen in the figure, communication preparation began in November of the previous year.

Using a variety of channels for communication is essential toward reaching the largest audience group. Table 3.2 provides a variety of channels for communicating about change initiatives within and external to the organization.

Recall the communication plan depicted in Figure 2.2. In that partially completed example, a variety of communication channels were used. The more channels utilized to get out the message initially and update on the change initiative, the more likely the organization is to engage the largest group of stakeholders.

In determining the channel to use, consider the emotions that need to be addressed regarding the change as well as the knowledge around why the change has to occur. Consider this scenario:

The CEO of a manufacturing firm that has been in business for over 40 years wants to makes changes to the production line. He is expecting that there will be some resistance primarily because a majority of the employees who work on the line have been with the company on average for 20 plus years and the status quo is appealing. Additionally, there is no pressing need to change the line, overall it is working well and is fairly efficient given the competition. His desire to change the line processes is because there has been significant growth, and he believes efficiencies can be improved upon. On the other hand, however, the CEO also has a strong working relationship with the employees, pays them over the average salary for their roles, and his treatment of them overall has actually kept the manufacturing firm union-free—unlike his competition. He takes

Communication Strategy

Change Project: Process improvement change initiative

Objective: To eliminate redundancies in processes that have developed through organic growth of three key areas within the organization.

Impacted Groups: Marketing, Sales, Operations
Number of impacted employees: 150 employees between all three departments involved
Other impacted individuals/employees/external parties: 2 vendors

Key Change Management Personnel

Name	Title	Primary Role
Jack Smith	CEO	Change Sponsor
Abigail Adams	SVP Marketing	Change Leader
Allan Johnson	SVP Sales	Change Leader
James Sanders	SVP Operations	Change Leader
John Pearson	Sr. Project Manager	Change Manager
Jeremy Fanland	Communication Supervisor	Communication Contact

Communication Tools and Channels

Tool/Channel	Primary Use of Tool/Channel
Department Meetings	Initial communication by each change leader to their department; prior to email to entire organization
All Staff Meetings	Communication to share information, ensure everyone hears same information, answer questions
Email	To communicate organization-wide to notify of change initiative and for regular update on change initiative
Surveys	To gather data to structure the change initiative from those most closely impacted
Focus Groups	Follow up to surveys; used to gather feedback throughout implementation
Internal Portal/Website	Another source for employees from throughout the organization to get information about the change initiative and share their thoughts/feedback

Internal Stakeholders and Information Requirements

Stakeholder Group	Information Requirements
Marketing	• Benefits of change • Training to be provided • Regular updates • Etc.
Sales	• Benefits of change • Training to be provided • How to communicate with customers • Regular updates • Etc.
Operations	• Benefits of change • Training to be provided • Regular updates • Etc.
All employees	• Initial communication about the *why* of the change • Regular updates about what is going on

Figure 3.6   Partially completed communication strategy

External Stakeholders and Information Requirements	
Stakeholder Group	**Information Requirements**
Vendor A	• What's happening and why • Impact from initiative • Etc.
Vendor B	• What's happening and why • Impact from initiative • Etc.
Customers	• Benefits to customers

Distribution of Information: Primary Point of Contact: Jeremy Fanland, Communication Supervisor

Requirements for Distributing Information

Communication Component	Due Date	Audience	Distribution Methods
Presentation	Dec xx, 20xx	All employees	At all staff meeting, via email
Internal portal	Dec xx, 20xx	All employees	Share portal access via email
Etc.			

Requirements for Information Gathering and Reporting

Information Input	Person(s) Responsible for Collecting and Reporting	Person(s) Responsible for Submitting Information	Due Date
Information for presentation	Marketing, Sales, Operations Administrative personnel	Marketing administrator	Dec xx, 20xx
Requirements for internal portal	Marketing, Sales, Operations	IT Group – Jason Simmons	Nov xx, 20xx
Etc.			

Issue Escalation Process:

Approval of all communications to reside with Communication Supervisor. All communications to flow through Communication Supervisor. Any changes to planned communications – whether reduction in communications or additional communications will be considered an "issue" and must be approved.

Issues regarding communications will follow this process:
- Meeting to be led by Communication Supervisor to understand and document issue
- Issues with impacts between $x and $x or of not more than 2 weeks in duration: decision rests with Communication Supervisor
- All other issues to be escalated to Change Sponsor and Change Leaders for decision
    - Include: 2 – 3 options to resolve; pros and cons of each option; Communication Supervisor to recommend option

Communication Plan Updates:

Communication plan to be evaluated every 3 months at a Change Team Monthly Meeting. Updates to occur as needed.

Revision History

Version Number	Date	Originator	Reason for Change
1.0	Nov xx, 20xx	Communication Supervisor	N/A

Figure 3.6 Continued

Table 3.2   Potential communication channels

Communicating Internally	Communicating Externally
• Internal portal/website	• E-mails
• E-mails	• Client portal
• Newsletters	• Client presentations
• Posters	• Focus groups
• Lunch and learn sessions	• Surveys
• Breakfast meetings	• Client meetings
• After hours events	
• Focus groups	
• Surveys	
• Informal conversations	
• Department meetings	
• Virtual presentations	

all of this into account and develops an initial communication plan as shown in Figure 3.7.

As can be seen in Figure 3.7, there is a number of planned communication points *prior* to the actual start of this change initiative. This enables engaging the employees on the manufacturing line in the change through ensuring understanding of the *why* of the change and gaining their participation and engagement in the change. Had the CEO had his first meeting explaining that the initiative was starting, the tone of his employees would be quite different. When organizations launch significant change initiatives before communicating about the change and allowing time for employees to provide input, the change is more likely to be met with resistance. Consider this scenario.

Paul is the head of finance for a national organization. He has been tasked by the CEO with restructuring his department to better support the growing organization. Paul takes two weeks and outlines a new structure and gets approval from the CEO for the proposed changes. He involves no one in his department in the initiative. In fact, not one of his employees knows about the assignment from the CEO. After the CEO approves his structure, Paul arranges for a meeting with his staff. He notifies them that the CEO has approved his reorganization of the department and that it will be effective within two weeks. All eight of Paul's employees react negatively. They remind him, first, that they are in the middle of closing the

Sender	Overall Timing (project phase)	Audience	Primary Focus	Message Content	Delivery Method	Date
CEO	Prior to start	All manufacturing line employees	Announce upcoming change – to be started sometime in mid- to late-July	Vision for change Explanation of change Request for assistance in making changes	Multiple meetings to cover all shifts Email follow up	Late June
CEO	Prior to start	All manufacturing line employees	Follow up and to secure interest in working on the initiative	Answer questions	One-on-one at line; meeting with employees in break room	Late June/ Early July
CEO	Prior to start	Subset of manufacturing line employees	Discuss potential areas of changes to production line	Discuss areas of possible change to improve efficiencies; set schedule for initiative	Meeting with all interested employees	Mid-July

Figure 3.7   Partially completed communication plan for manufacturing line change initiative

> *books for the year. Second, his proposed structure—even with just a glance at it—will not work given a number of factors. Paul is shocked at the reaction.*

Let's stop here and explore this particular scenario further. This change initiative has already failed. Because finance department employees were not engaged early on—either in learning about the impending change or in helping to restructure the new organization—they have not accepted it. And, they have raised at least two important red flags: the books are being closed for the year and the proposed restructuring will not work. Had Paul, the head of the group, engaged his employees early on when he was initially tasked with this assignment by the CEO, he would have gotten the buy-in he needed for it to be successful. Now, Paul will have to step back and get his employees involved. But first, he'll have to get past the anger that exists because of the initial noninvolvement of the group in a change that directly impacts them and their work. This anger will create resistance to the initiative; even if it is in the best interests of the organization and the employees in the group. Let's consider the same scenario, but in this case Paul has engaged his team.

> *Paul has been tasked by the CEO with restructuring his department to better support the growing organization. Given that it is only a month before the end of the fiscal year, Paul knows that his team will be focused on closing the books and he needs their involvement. After all, they are the ones who are doing the job each day. Paul notifies the CEO that he will submit a proposed restructuring of the department after the first of the year, when the books have been closed. During a regularly scheduled team meeting, Paul notifies his group that they have been tasked with restructuring the department to support the significant growth of the organization. He tells them that the priority is closing the books and that at the start of the year he'll get the team together for an off-site meeting to discuss how to proceed. At that time he will share more information. He also tells them that no one will lose their jobs and, in fact, the group will be growing. At the start of the year, Paul gets the group together for an off-site meeting to discuss the change initiative ahead of them. He shares his vision for the change initiative and expresses excitement about the opportunities for the group to be a bit more focused on strategy. He also tells them they will be the ones leading the initiative and developing the new structure. The team is excited and begins to discuss a variety of options. Paul has engaged his team in a significant change effort.*

In this scenario, Paul's team is excited and engaged in the initiative. They will become champions of the initiative because they are directly involved in crafting how they will get their work done. They have been given ownership of the effort by their manager.

Getting a Cross-Functional Group Involved

In Chapter 2, a Stakeholder Support Committee was mentioned. Let's expand more on the concept of a Stakeholder Support Committee and its value in implementing organizational change initiatives. Whenever an organizational change initiative:

- is significant and complex;
- has multiple locations involved; and/or
- has broad impact among the employee base…

the use of a Stakeholder Support Committee will be beneficial. Such a committee will take some of the responsibility over communications off the shoulders of leadership as well as the project manager who is leading the change; and it engages more individuals in the change overall. Recall the statistic shared in Chapter 1—that 80% of employees accept change when that change is supported by influential nonleaders in the organization. The Stakeholder Support Committee is comprised of the organization's nonleaders who will help to influence others to embrace the change and see the good in the change. Table 3.3 provides some characteristics of strong Stakeholder Support Committees.

Building a Stakeholder Support Committee is a great way to engage a larger group of individuals in a large, complex organizational change initiative. The benefits of such committees include:

- Enabling the change leader to stay closer to the employees impacted by the change
- Supporting sharing of information from the change leader to all employees and back up the ladder—from employees to the change leader
- Serving as a pilot group to test any changes before they *go live*
- Providing feedback throughout the change initiative

Consider the Stakeholder Support Committee as another communication channel to share information about the change initiative. In fact, this may be the most important communication channel, as it is made up of individual employees (nonleaders) sharing their excitement about the change with other nonleaders (their peers).

Table 3.3   Characteristics of strong Stakeholder Support Committees

• Comprised of members representing all impacted groups within the organization • Comprised of members from all levels (not solely management staff) • Provided a point of contact on the leadership team to remove barriers and address issues that need leadership attention • Provided time during regular working hours to engage in committee activities	• Provided the autonomy to interact throughout the organization and engage others in change • Provided the tools needed to be successful in engaging employees • Provided a budget for food and beverage for meetings, posters promoting the change, give-aways (e.g., buttons) and for other items to help in promoting the change and engaging employees

Sharing the Communication Strategy and Plan

Sharing the communication strategy and communication plan for the change creates feedback from those participating in implementing the change. Sharing of this information is of particular importance if leadership is relying on the efforts of a Stakeholder Support Committee to keep employees throughout the organization informed about and engaged in the change.

While the goal with a Stakeholder Support Committee is to enable them to communicate about the change initiative, there is still a need to manage those communications. While leadership wants to ensure regular communications, there must be a plan to ensure that those communications are not in conflict and are not so much that those on the receiving end are overwhelmed by the change initiative.

> *One VP of Communication for a global organization undergoing a complex change initiative developed a communication plan that alternated change project updates between the communication department and the Stakeholder Support Committee. This was primarily done through information about the project sent to all employees on a biweekly basis via e-mail and posted on an internal website. On alternating weeks, the Stakeholder Support Committee would then focus on following up on the communication that was sent by the communication department. This enabled individual employees to internalize the information they received on the change project and then to discuss it further with their peers. This was an effective way of keeping the organization engaged and ensuring just the right amount of communication. It also enabled the project manager to focus on ensuring the change project kept moving forward toward a successful conclusion.*

Ensuring Regular and Consistent Communications

Developing a strategy and plan for communicating the change, including using a Stakeholder Support Committee to help communicate during complex change initiatives, stimulates regular and consistent communications. Organizations that fail to think strategically about communications during change will fail to engage employees early on and fail to keep them engaged. Consider the preparation one national organization conducted *prior* to launching their first strategic change initiative:

A national consulting firm was launching its first strategic, complex change initiative that would impact all functional areas. Two months prior to officially launching the initiative, the CEO facilitated a meeting with all senior leadership to achieve the following:

- *Ensure a shared vision of the change and support at the senior leadership level*
- *Develop a communication strategy to enable effective and sufficient communications about the change*
- *Craft an initial communication to be shared during staff meetings and via e-mail from the CEO that focuses on the why, what, who, when, and where of the change*

This time spent preparing for how the organization will communicate about the change enabled a successful change initiative overall for the organization. It is a best practice that continues to be used within the organization today—even for smaller change initiatives.

The most effective way to ensure regular and consistent communications is to follow the five steps outlined in Figure 3.8.

SETTING A VISION FOR CHANGE AND COMMUNICATING THAT VISION

Proactive change is launched because a leader has a vision for the future success of the organization. Employees who get behind change do so because they can understand and *see* the vision. It is expressed in terms that engage employees and get them excited about the possibilities for the future. While sharing that vision is essential; it is also important to ensure an

Step	Do this...
1	Develop a communication strategy and communication plan for the executive
2	Have a variety of channels for communication to reach the broadest audience
3	Ensure specific roles and responsibilities for communications
4	Share the strategy and plan with everyone responsible for communications
5	Regularly evaluate and update as necessary the communication plan throughout the change initiative

Figure 3.8 Five steps to ensure regular and consistent communications

understanding of the *why* of the change. When developing the vision for a change, leaders need to be sure that vision will produce answers to the questions from employees that are shown in Figure 3.9.

Consider this vision for a change initiative:

> *A nonprofit was launching a significant change initiative within the organization. The Executive Director shared his vision with employees and volunteers: "This change will enable the organization as a whole, our employees and our volunteers, to do the job they want to do—to improve the lives of those around us by providing them the support they need to get educated and find jobs to support their families and homes to enable them to be safe. Imagine a future where every adult has the ability to support their family and every child feels safe... that is our vision and we need your help to get there!"*

This vision of the Executive Director of a nonprofit organization was presented to a room of over 60 individuals who worked for and supported the organization (the majority of whom were volunteers). Every individual walked out of that room excited about the future. In following up with some of the volunteers afterward, here is what we heard:

- They were excited about the vision
- They were thrilled that the change would enable them to focus on their reason for joining the organization

Will this question be answered when the vision is shared…	Yes/No
Does the vision for the change make sense given the current culture of the organization?	
Does the employee understand the vision?	
Can the employee explain the vision to another individual in just a few sentences?	
Does the vision translate to the good of the change for the individual as well as the organization?	
Does the vision motivate and engage the employee?	
Does the employee believe the vision will enable him/her to better perform his/her role?	
Does the employee understand what he/she needs to do personally to achieve that vision?	

Figure 3.9   Questions to be answered in sharing the vision with employees

- They felt the change was well-thought-out and the organization was well-positioned to implement the change
- They felt that the Executive Director was excited about the future which, in turn, got them excited about the future

While it is important to have a number of individuals who are outside of the leadership level communicating and talking about change when it is being implemented within the organization, it is absolutely essential that sharing of the initial vision and kicking off the change initiative is done by an executive in the organization.

TELLING STORIES AND PLANNING FOR THE FUTURE

There are two ways to discuss change within the organization—a focus on the rational side of change and a focus on the emotional side of change. Figure 3.10 provides a description of each side of change.

While both sides of the discussion are important in moving forward with change, the emotional side is key to engaging employees in change. Change is a very personal matter for individuals. What drives one individual to change may not drive another. Therefore, while sharing information that covers both the *rational* side as well as the *emotional* side, it is important to put more emphasis on the *emotional* side when looking to engage employees in change. The emotional side is what will help employees to accept and embrace change more than the rational side.

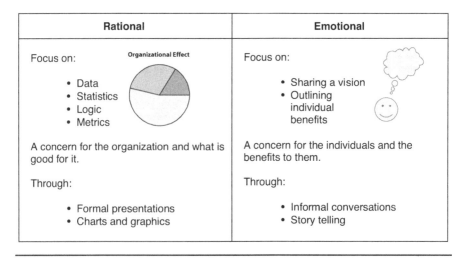

Rational	Emotional
Focus on: Organizational Effect • Data • Statistics • Logic • Metrics	Focus on: • Sharing a vision • Outlining individual benefits
A concern for the organization and what is good for it.	A concern for the individuals and the benefits to them.
Through: • Formal presentations • Charts and graphics	Through: • Informal conversations • Story telling

Figure 3.10   Two ways to discuss change

A great way to enable employees to make an emotional connection with a proposed change is to share a story about the future of the organization. Take the vision and show how it will look in the future through telling a story. Consider the story shared about the nonprofit organization. Here is another way the Executive Director might present the vision to employees and volunteers:

> *It is 10 years in the future. Every individual and family who has needed our help has received it. We can claim that we have enabled individuals to receive education to get better jobs. Because they have been able to get better jobs, families have sufficient food and are in homes in safe neighborhoods. In our area of influence, no child is hungry and no family lives on the street or in their car. Congratulations to all of you for your efforts in making the world a better place!*

Now, imagine yourself as one of their volunteers. This is exciting! This is compelling! This engages people in *wanting* to embrace and move forward with doing what they need to do to make this story a fact. Such storytelling is far more effective than simply sharing information, such as how many families the organization supports today and how many it wants to support in the future. Yes, that is important of course, but the story of a possible future is much more exciting.

What's in It for Me?

When employees first hear of change, the first thought that goes through their mind is—*What's in it for me?* What they are asking is:

- Why should I change?
- What will I get out of it?
- What is the impact on me?
- Will my job change?
- Will I have the skills I need to be successful?
- When do I have to make changes and how will I be supported?

This is why focusing on the emotional side of change is essential. Individual employees will be able to answer the question—*What's in it for me?*—when they understand the change from their own perspective. This means understanding the value and benefit of the change for them as individuals.

> *Jennifer, the head of a regional company, spoke with her employees about the need to change how they were solving problems to better compete in an increasingly competitive environment for their*

products. She could have focused on the benefit to the organization in changing; specifically that they, as a company, would be able to grow and thrive and employees would have jobs. But rather, she focused on the individual employees by sharing that this change—if successful—would enable employees to focus on creating new products rather than fixing problems with the current products. They would be able to spend more time brainstorming and being innovative—the reason that they joined the organization in the first place. Jennifer shared that she envisioned an organization where each employee would have the time to be creative in their roles. The need to react would be reduced dramatically and employees would be more proactive in their roles.

Refer back to the discussion around the Five Ws earlier in this chapter. When considering the response to these Five Ws in discussing change with employees, consider how to incorporate specific information to enable employees to understand the value and benefit of the change for them personally. Figure 3.11 provides a strategy for addressing the Five Ws with a focus on the individual employee.

In particular, the *why* and the *when* components are focused on the individual employee. The change is being explained from the perspective of allowing a greater focus on engaging customers and growing the customer base. From a sales perspective, the ability to grow the customer base would mean more money in their pockets, and from a customer service perspective it means being able to be more proactive with customers and therefore, develop stronger relationships. The change is also being launched during a

Change Initiative: Change Processes and Procedures in Setting Up New Customers	
WHO	All customer service and sales employees
WHAT	Refine processes and procedures used in setting up new customers
WHY	To enable for less of a focus on administrative tasks and more focus on engaging customers and growing the customer base
WHEN	During the summer months when there is less pressure on customer service and sales employees
WHERE	Across the entire organization; all customer and sales employees

Figure 3.11   The Five Ws with a focus on the individual employee

slower period for both groups of employees. Rather than launching the change during what may be a busy period—for example when salespeople are trying to make their sales quotas—it is being launched during a slower period so employees can focus appropriately.

If the organization was looking at this from an organizational focus, the *why* may be to drive increased profitability for the organization and the *when* may be immediately without a concern for what else is going on for employees.

While not every employee, regardless of what tactic the organizational leadership takes, will look at change from a positive perspective; leadership is more likely to engage those negative employees sooner rather than later if they follow these best practices. While it may seem that such practices will take significant effort and may delay the launch of a change initiative, research bears out that those organizations that focus on communicating early on and in the right way about change to engage employees are much more likely to be successful overall in their change initiatives. Over time, lessons learned have shown that organizations that focus on change from a proactive perspective with a focus on the value to the organization *as well as* the individual employees are more likely to increase the engagement of their employees in change initiatives in the future.

One of my clients needed to be convinced that promoting the need for change before actually starting the change initiative would be more successful for him than his usual practice of launching change without sharing the why of the change. I asked—actually, I pleaded—that he give me just one month to talk about the need for change before he officially kicked off the change initiative. The client reluctantly agreed to the one month. At the end of the month, after engaging employees in a variety of ways about the why of the change and the value to the employees, the initiative was launched. Six months later, checking in on progress, the client commented that his employees were really engaged in the change and, in fact, offered a number of suggestions that were making for a fantastic end result. Each time a change is launched within this client organization, I go in to help leadership socialize and engage employees in shaping that change.

4

BUILDING CHANGE
CAPABILITY WITHIN THE
ORGANIZATION

"To improve is to change; to be perfect is to change often."
Winston S. Churchill

The most successful organizations are forever changing. Change becomes the norm in these organizations. Change is seen as a positive experience—an opportunity to grow, strengthen, and improve both the organization and its workforce.

For one consulting firm, the need to reinvent itself was the catalyst for change in the organization. This need to change came from a threat from other firms that only a few years ago were not even in the rear-view mirror of the company. New senior leadership recognized the need to drive change—and to do so quickly—in order to respond to the competition and ensure the survival of the organization. Fortunately, many employees already knew this was necessary and were engaged in the change. A lesson learned to the organization—which they took to heart—was to regularly evaluate the organization against its competition to ensure that any future change would be more proactive than reactive.

The capability to change within the organization is essential for any organization trying to achieve strategic initiatives. All complex and strategic initiatives launched within the organization require a cultural change and a change to the behavior of the individuals working in the organization. When an organization launches a project that will change how the work gets done, this requires employees to change their own behavior as to how they approach the work. The Economist Intelligence Unit research report, *Why Good Strategies Fail: Lessons for the C-Suite* (July, 2013), noted that an organization's lack of change management skills is a primary reason for why strategic initiatives fail. Organizations that succeed at strategic initiatives understand that change capability is essential and can check off the items depicted in Figure 4.1.

The ability, even the willingness, to continuously change, however, is more easily said than done. Building a change-capable organization takes time, effort, and leadership willing to support change. This chapter will focus on how to build a capability around change within the workplace.

THE VALUE OF ORGANIZATIONAL CHANGE

Organizational change is essential if an organization is to grow, prosper, and remain competitive over the long term. No organization can sustain without change.

> *All Company Training noticed that, year after year, their ability to increase their customer base was diminishing. Without consistent growth in their customer base, All Company Training would eventually lose revenue and see decreased profitability. The problem was that All Company Training's marketing processes encouraged retaining current customers and not getting new customers.*

The following is in place in order to ensure a change-capable organization…
A structured, tested strategic project management approach to how change is managed
Leadership that is trusted within the organization
Leadership capable of leading and championing change
Mid-level management that engages in and champions change
Plans in place to engage employees in change
Plans in place to ensure change "sticks"
A variety of communication channels to share messages of change
A positive perspective of change
Project management capability to implement change initiatives
A mandate for continuous change

Figure 4.1   Checklist: is the organization change-capable?

*Additionally, sales were commissioned more on retention of custom-
ers than on bringing a new customer on board. All Company Training's leadership needed to launch a change initiative that would accomplish the strategic goal of increasing the customer base by 25%
over the next three years.*

The more complex a change initiative is, the more likely that it requires
other changes to be implemented first. All Company Training's strategic
goal—and primary change project—is to increase the customer base by
25% over the next three years. This change, however, will require additional changes to take place before it can be accomplished. This includes:

- Changes in processes that are related to customer acquisition and
  customer service
- Changes in sales compensation and bonus structures to focus sales
  on pursuing new customers
- Changes to marketing strategy to focus on new customer acquisition
- Restructuring the sales organization so that some employees are focused on new customer acquisition and others are focused on customer retention

All of this requires engaging each impacted employee in the change. By
breaking down this change to doing the smaller necessary changes prior
toward accomplishing the larger change, the organization creates increased
buy-in by employees and is more likely to see an overall success in meeting their goal of increasing the customer base by 25% over the next three
years. All Company Training cannot possibly simply achieve an increase in
their customer base without *first* determining what needs to change in the
organization to help facilitate their strategic goal.

Organizational change should promote a variety of benefits, including:

- Staying ahead of the competition
- Stimulating steady and consistent growth
- Innovation in products and services
- Improved problem solving

Making Change Part of the Strategy

As mentioned previously, change needs to be aligned to the organizational
strategy. You may recall that strategic initiatives *require* change to occur
in order to be successful. As a best practice, leadership might think about
doing the following:

In order to ensure that change is considered from a strategic point of view, one Chief Executive Officer (CEO) and his leadership team consider what else has to change for a strategic goal to be successful. The members of the senior leadership team meet annually, off-site, to discuss their strategy for the upcoming year. Here is a glimpse into what occurred at a recent strategy session.

Two strategic goals were planned for the upcoming fiscal year:

- *Open a global office*
- *Move into a new line of business*

Prior to launching these two key strategic initiatives, however, the CEO and his leadership team considered the organization as it was today and determined that to be successful, a number of changes would have to take place first:

- *Restructuring of the organization*
- *Changes to current processes*
- *Changes to the current rewards system*
- *Changes to metrics used to run the business*

These changes had to occur before the two key strategic goals could be implemented if the organization was going to be successful in opening a global office and starting a new line of business. Additionally, these changes outlined in the list above required culture change within the organization.

Successful organizations review strategic initiatives not just from the perspective of how it will drive more revenue, increase profitability, and reduce expenses (to name just a few benefits of strategic goals), but also from the perspective of what change is necessary within the organization to support the strategic initiative and how that change will impact the culture of the organization.

Consider an organization where productivity is down. Rather than just demanding increased productivity, the smart organizational leader will set a strategic goal to improve productivity by a specific percentage within a specific period of time. This then leads the organization to launching a change initiative to improve productivity. Launching this strategic goal as a change initiative in the organization focuses the organization on looking at *what needs to change* within the organization to improve productivity over the long term and what barriers might exist toward increasing productivity. The change initiative may bring to the surface the following issues to be addressed as part of the overall change initiative of improving productivity:

- Refined processes
- Reorganization of particular departments
- Implementation of new technology
- Training of staff

By looking at this goal as a change initiative, and applying change management best practices to implementing the initiative, the organization is more likely to find a long-term solution to the problem of decreased productivity.

ASSESSING THE READINESS FOR CHANGE

Change is difficult. This statement has been made many times since Chapter 1 and will continue to be emphasized throughout this book. When leadership realizes the difficulty in implementing transformational, strategic change initiatives, they are more likely to invest in them sufficiently to ensure they are successful. The biggest barrier to success is leadership who believed that change was easy or a no-brainer.

Some organizational leaders also forget that being ready for one change does not mean an organization is ready for another. However, there are some things an organization may do to increase their change readiness for all changes over time, thereby reducing the effort of implementing large change initiatives within the organization. Figure 4.2 provides a number of these key considerations to ensure change readiness in the organization.

To avoid confusion, let's contrast Figure 4.2 with Figure 2.6. Figure 2.6 is focused on being ready for a *specific* change in the organization. Figure 4.2, on the other hand, is focused on an overall preparedness, or readiness, for regular and consistent change. Can your organization check off many of the items listed in Figure 4.2? If not, consider what might be done to increase the readiness for change within the organization, starting at the leadership level and moving down through to all levels within the organization.

Just as organizations change over time—they grow, react to new competition, launch new product lines, etc.—the way they implement and manage change will also have to be adjusted. Consider this story:

> *The most significant change initiative I have experienced in my 25-year career was in the early nineties in an international*

Ensuring Change Readiness	
The organization....	
	functions well in change
	invests budget monies in implementing change
	invests and supports employees in implementing change
	ensures technology supports change initiatives
	ensures there are sufficient resources to launch change initiatives
	has trained its leaders and employees in managing change
	knows the value in engaging employees in change early on
	takes time to identify and break down resistance to change
	acknowledges barriers to change and takes time to remove them
	identifies non-leader champions of change and empowers them to lead change
	ensures those involved in change are enabled to work on the change initiative away from their day-to-day responsibilities
	evaluates what in particular must happen to sustain a specific change over time (ensure the change "sticks")
	has processes and procedures in place and uses project management best practices to manage change regardless of its size and complexity

Figure 4.2   Ensuring change readiness

telecommunications company in North America. I was working in a division with two thousand people. Half of the employees worked in the production of sophisticated printed circuit boards and associated equipment; the other half in offices supporting the sales, marketing, production, and distribution of this equipment. The division had just received its ISO-9001 certification a few months earlier. In this growing company, a groundbreaking change initiative was created by Corporate called the Excellence Initiative. This initiative had as a goal to implement the latest thinking in change management (CM) by creating Excellence Teams to focus on solving some intractable issues. The Excellence Teams were, by design, cross-functional and meant to take some time away from the day-to-day, for mandates varying between three months to a year, to implement concrete improvement in either processes, tools, or other productivity related activities. I was very impressed with the breadth and depth of this initiative. Several employees, chosen from different functions, were trained for more than two weeks, full time, on the Excellence Initiative itself, on its principles and approach, and some were trained on group facilitation skills and techniques. This first wave of trainees also received "Train the trainer" classes so that they could come back to the workplace and train Excellence Team leaders. In the span of roughly three months, 30 Excellence Teams were formed, each containing between five and eight people, with the part-time mandate of working on Excellence Team assignments in parallel with their current jobs. The initiative yielded significant change, the large majority of it beneficial. It was a worthwhile effort to bring about needed change through strong focus. In hindsight, the main reasons in my mind for this initiative's success was its top-down mandate from the corporate office, the strong backing of local executives who wanted to show what could be done and benefit from the needed changes, and the commitment of training a critical mass of employees in the philosophy and tools of the initiative, thus obtaining broad buy-in.

Let's pause here in this story. The organization depicted here saw change as a long-term strategy. Change was made a strategy through the launch of Excellence Teams. They took the time needed to engage their employees in change—ensuring they had the skills, knowledge, tools, and technology necessary to be successful in leading and implementing change throughout the organization. Change was to be a way of life for the organization and its employees. Let's go back to the story shared by a friend and colleague:

In the first part of this story, I relayed an impressively successful CM endeavor in the early nineties adopted by a major telecommunications company in North America. With 30 teams set up and trained on this Excellence initiative within three months, with several teams yielding concrete and beneficial results and then disbanding once the job was done, the overall initiative was a huge success. Then, why did the initiative die? A mere 14 months after investing time and money to train dozens of facilitators and put in place more than 30 teams with temporary mandates to improve business productivity, not one Excellence Team remained. A head scratcher isn't it? If the initiative was so successful, why did it not continue beyond a very productive 14 months? There are several reasons that could explain this outcome. Had we run out of issues to solve? Probably not. Had we run out of processes to improve? That would go against the adage of continuous improvement. Had we run out of steam? Probably. I believe that after spending a lot of energy and focus toward improving on and solving many issues across the division and the company, employees "tired" of the initiative's high focus. It wasn't the "latest thing" anymore. The lesson learned here is that the approach and strategies in which we are to implement change in organizations are themselves subject to change. The only thing that does not change is the necessity for change. Implementing CM is very difficult for the simple reason that it is highly contextual and requires some kind of "newness" in its approach or benefit. There's a reason that CM is best implemented with a systematic approach, a sensitization of the stakeholders, and guiding principles—it's hard!

CREATING AN ENVIRONMENT FOR CHANGE

Figures 4.1 and 4.2 assist in preparing the organization for change and ensuring change readiness—covering many key areas including:

- Ensuring leader sponsorship of the change
- Engaging employees in change
- Implementing the change using strategic project management practices and processes
- Sustaining the change over time

Figure 2.4 provided a four-step process to engage the organization in change. In particular, the first step was focused on *creating an environment that welcomes and embraces change*. Let's expand on creating that environment. Figure 4.3 provides one potential flow for ensuring an environment that will embrace change.

Table 4.1 explains each of these steps in more detail, providing a brief description on what would occur in each step.

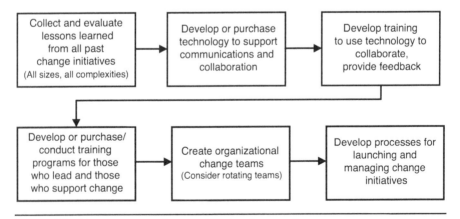

Figure 4.3 Ensuring an environment that embraces change

Table 4.1   Creating an environment that embraces change

Step	What Is the Step	How to Proceed
1	Collect and evaluate lessons learned from all past change initiatives	• Review successful and failed past change initiatives • Explore why initiatives were successful or why they failed • Develop a plan to address failures from past change initiatives (how will the next change initiative be different)
2	Develop or purchase technology to support communications and collaboration	• Determine tools and technology needed for managing change initiatives regardless of complexity—e.g., collaboration portal to share information, e-Newsletters to communicate about change, e-mail templates, etc.
3	Develop training to use technology to collaborate and provide feedback	• Develop training modules to enable for more effective use of tools and technology (e.g., webinars to use collaboration portal, component of onboarding program for new hires, etc.)
4	Develop or purchase/ conduct training programs for those who lead and those who support change	• Develop and/or purchase training for change leaders as well as for those supporting change initiatives (provide a variety of formats for "just-in-time" learning options)
5	Create organizational change teams (rotating teams)	• Develop a process for assigning or nominating individuals to serve on organizational change teams • Ensure organizational change teams are comprised of a senior leader (change sponsor) as well as individuals representing all functions in the organization • Develop metrics as well as performance criteria for evaluating success of change teams • Create a charter for the change team • Develop norms, processes, and procedures for working as a change team
6	Develop processes for launching and managing change initiatives	• Determine project management best practices processes as well as documentation for launching and managing change initiatives • Determine how progress will be reported, how information will be shared, how requirements will be gathered, etc.

Let's look at these steps from the perspective of a case study. Consider this situation of a CEO of a small business with a goal of significant national growth over the next eight to ten years:

Melanie's business currently employs 100 individuals within its home state. Growth has been consistent, and Melanie is now ready to open up a number of offices nationally with a goal of 500+ employees across the United States. While this is the largest change the organization has ever undertaken, there have been other smaller change initiatives over the years that have been less than successful. Melanie, after conversations with 25 employees who have been with her since the founding of the business (Step 1 in Table 4.1), learned the following important lessons on previous failings:

- *There was limited communication about the change*
- *Employees did not understand why the change was happening*
- *No one was really leading the change*

This information will be used to frame this upcoming change plus all future change initiatives. Melanie assigns a project manager to oversee the change project overall. As a start, she also tasks the information technology (IT) department with sourcing technology to enable collaboration for this change initiative with the goal that as the organization grows, this collaboration tool will be even more essential to engage employees (Step 2, Table 4.1). Additionally, Melanie, in collaboration with the assigned project manager and along with employees from the Communication Department, developed a strategic communication plan. The first communication would be focused on the following content and delivered during a half day all-hands meeting:

- *A vision for the change*
- *Explanation of the change (the "why")*
- *What to expect*
- *How technology will be used (collaboration portal)*
- *Benefits to the organization and employees*
- *Lessons learned from past change initiatives and how this one will be different*
- *Notice of project start within two months as well as the name of the project manager assigned to lead the initiative*

Melanie led the meeting but ensured each of her department heads were supportive and champions of the project as they were tasked with holding follow-up meetings with each of their staff to answer any additional questions.

Department heads also shared in follow-up sessions that training would be provided to all employees on how to use the collaboration portal since it would be an ongoing technology within the organization (Table 4.1, Step 3) and that those selected to serve on the project would be provided training regarding managing change and all other employees would be provided training pertaining to supporting change (Table 4.1, Step 4). Of particular importance to engage employees, department heads also noted that at some point in time every employee would be able to serve on a change team that would be a rotating change team with a 12-month charter (Table 4.1, Step 5). This was particularly valuable to employees in an organization that, so far, had limited career path opportunities. Being selected to serve on a change team was seen as a great professional development benefit for employees. (It was also viewed as a "what's in it for me" for employees.) The change team selected for this first initiative was comprised of six individuals from throughout the organization (representing all functions) led by the project manager who was assigned early on in the initiative. This team, with the support of leadership, was responsible for developing the processes and procedures to be used for managing all change initiatives as they moved forward (Table 4.1, Step 6).

While this is a simple example, it provides an overview of how an organization may apply the steps to create an environment that enables people to embrace change. In this case study example, the CEO, Melanie, is establishing best practices around not just this one change initiative, but also future change initiatives.

Organizational Culture Matters

Earlier chapters shared the need to understand the challenges of individuals accepting change as well as the impact of the change initiative on individuals. People and culture are linked and any change initiative must consider both of these elements. The culture of the organization must be considered when a change initiative is launched. This is because change is very much impacted by the current organizational culture. The culture of the organization is rolled up in its values, mission, and strategic goals. It is also defined by the types of employees who are hired and how they are rewarded as well as the way that work gets done and decisions get made. Culture is a shared understanding of the organization and an understanding that the perspectives, interests, beliefs, and ideas of every individual within the organization shapes and feeds its culture. It is not always easy to define culture. Add to that the fact that the larger the organization, the more likely there are subcultures that exist. In fact, subcultures can exist in organizations with less than 200 employees. Subcultures arise within departments, divisions, and even within particular working teams within a larger team environment.

When a change leader hears from employees, "We don't do things that way around here," this is an indicator that the change is not aligned with the current organizational culture. The culture of the organization has a *way of getting things done* and if a change initiative is launched that will dramatically change how things get done, it will be resisted by the culture of the organization. Let's look at a very simple example of launching a change without considering the culture of the organization:

> *Nadia launched a change initiative with a focus on how she would reward her team. She was changing the reward system in place for her customer service group by rewarding them with bonuses and promotional opportunities for collaborating with each other to solve customer issues. She wanted them to start working more collaboratively as a team rather than individually in managing their customers. She developed a comprehensive approach to how her customer service team would be rewarded with bonuses for team work over individual efforts. Nadia got approval from her manager to present it to the team and make it effective at the start of the new fiscal year. Nadia forgot just one thing. Her team's performance review was focused on individual customer service response times in solving customer issues. All metrics—handed down from leadership—were focused on individual key performance indicators. The culture of her customer service group, and many other areas within the business, was focused on individual efforts, not team work. Before Nadia could even consider changing how she awarded bonuses, she would need to work to change the organizational culture that was more focused on the individual than the team. This required changes in the performance management and reward system as well as in the structuring of the customer service group. She also needed to move her employees from being focused on themselves to being concerned about the team as a whole.*

Too often in change initiatives, culture is not addressed or is considered an afterthought, as in the simple example provided above. Successful change requires understanding the current culture of the organization and how that culture will either support or negatively impact the change. This doesn't mean that the culture can't be adapted to support the change, but it will take time and effort. If the culture will negatively impact the change—or does not support the change—it will have to be examined *before* the change can take place, otherwise the change will not be successful. Culture will win out over change every time.

> *Antonio was an executive in a global training company. The company's culture had a number of strengths, including collaboration among departments, sharing of workloads, and shared team bonuses. Since their inception, the organization has focused on a collaborative culture where only those employees who considered the team over the individual could be successful and would be rewarded. Given the significant growth the company was undergoing, and the increased competition for their training services, Antonio wanted to change things up a bit to drive his sales team to accomplish more. He decided that he wanted to change the focus for sales to the individual over the team, thereby increasing competition among sales employees. This would entail assigning specific client accounts to individuals and attaching bonuses and merit increases to individual goals, not team goals. However, Antonio also realized this could not happen overnight. Every sales employee who had been hired was hired because they were team-focused, they did not have the drive that you would see from salespeople who were focused on themselves and how much they could earn. The collaborative team structure was enforced through team goals and team bonuses. Regions were assigned to teams of salespeople, not individuals. Antonio realized that before he could realize his goal of individually focused sales with a strong drive to succeed, he needed to focus on changing the culture within the sales group.*

Before a change initiative is launched, leadership should consider whether the change is aligned to the current organizational culture. For example, does the change enable the employees to keep working as they normally do or as close to normal as possible? Or, does the change require the employees to dramatically change how the work gets done? Either can be accomplished, but the latter requires a change in the culture—the very fabric—of the organization. The former is a faster change as it is more likely to be adopted within the organization since it requires very little behavioral change on the part of the employees. When considering whether the culture supports the change desired, consider both formal components as well as informal components of the organization, as shown in Table 4.2.

Most organizational leaders consider the impact of change on the formal components of the organization, but, unfortunately, neglect the informal

Table 4.2   Formal and informal components of the organization

Formal Components	Informal Components
• Processes and procedures	• Employee and manager connections
• Reporting structures	• Cross-functional networks
• Defined roles and responsibilities	• Ad hoc gatherings of employees
• Decision-making rules	• Peer-to-peer collaboration and interactions
• Formal programs such as training and organizational development	• Communities of interest/social networks
• Performance management systems	• Beliefs and assumptions of employees
• Compensation, bonus, and reward systems	• Informal leaders (influential non-leaders)
• Internal communication processes	
• Formal committees within the organization	

components. Both components comprise the culture of the organization and both must be considered and aligned to any change initiative launched.

> *Selena was the CEO of a global organization. She was launching a large change initiative within the organization that would impact all functions, across all locations. In visits to all locations over a three-month period, Selena noted a number of informal groups that had risen up within the organization over the last few years. This included:*
>
> > • *A group of millennials who gathered together weekly to have lunch*
> > • *A group of women who gathered monthly to support each other in their roles through brainstorming and problem solving*
>
> *Additionally, Selena noticed that employees regularly reached out to each other to share ideas and discuss problems. She also saw a number of instances where managers were mentoring employees both within and external to their department. None of these gatherings were part of the formal structure of the organization, but certainly were an important component of the organization and the work getting done. Selena knew that in order to ensure her vision for the change could be achieved, she needed to engage these informal groups in helping her to shape and lead the change she envisioned.*

Understanding both formal *and* informal components of the organization is essential to achieving the success of the change initiative. While formal components are easily identified by leadership, informal components are not necessarily easily discovered. In our previous example, Selena, by visiting each of the locations, noticed these informal components through observation. Had she not visited each location, there is no guarantee these informal components would have been discovered, which would have been detrimental to her vision for change.

In addition to understanding and examining both formal and informal components of the organization, launching strategic, transformational change requires understanding the organizational culture strengths and weaknesses and whether or not the change desired is in alignment with the culture.

To ensure that change will *stick*, organizations must determine the path to get from the current culture (which may not support the desired transformational change) to the new, desired culture. Figure 4.4 provides a high-level example of how to understand a gap between the current culture and the envisioned desired culture—and then develop a plan to close that gap.

Let's look at Figure 4.4 in more detail. The organization in this example is not one that would normally support innovation, though increased innovation was a desire of the new CEO and the new Board of Directors. The new CEO was an individual with an entrepreneurial spirit; the last CEO had a very hierarchical approach to running the organization and decisions were made *only* by him and his senior leadership team. Additionally, the Board of Directors had recently turned over and new members

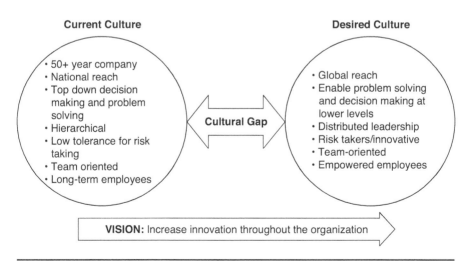

Figure 4.4   Current culture → new culture

were all individuals who had either run their global organizations or had come from investment firms—all agreed that innovation was the only way the organization would continue to survive in a highly competitive marketplace. While this certainly provided the CEO the support he needed to move forward with the vision, he also needed the support of the employees. As noted, the organization was hierarchical in structure. There was a low tolerance for risk and there were many long-term employees who had *settled* into the status quo. There was a lack of accountability at the lower levels of the organization with most employees leaving all problem solving and decision making with their direct manager; who, in turn, pushed problem solving and decision making up the ladder to senior leadership. This tended to create a bottleneck in resolving problems and responding to customers. The organization did tend to be team-oriented, which is one area where the culture would support the proposed change of becoming more innovative.

The new leader and Board understood that to be innovative required more than just demanding innovation and setting up brainstorming sessions—it required a change in the culture of the organization. The organization could not get from where it was to where it needed to be without changing its culture. Figure 4.5 provides some details as to specific activities that were undertaken to move the organization to the desired culture in order to achieve the ultimate strategic goal of being more innovative.

Figure 4.5   Moving to the new culture

As can be seen in Figure 4.5, the organization determined that areas of focus within the culture to achieve the vision included:

- Restructuring the organization under a new model that called for distributed leadership rather than top down control
- Advocating risk taking, which would lead to innovation, via refined processes and procedures
- Encouraging more leadership from employees through a refined performance management system including rewards and promotional paths
- Training employees in decision making and problem solving

There was also a goal to grow globally as many of the competitors had done, but that would be a secondary effort as the organization would need to have restructured processes and procedures refined prior to considering global growth.

While this is a high-level example, it provides insight into how an organization needs to consider their culture and whether or not there is alignment with strategic goals. Trying to move forward with increasing innovation within an organization that is not structured for innovation and employees who are not empowered nor rewarded to be innovative would not lead innovation. Rather, such an effort would fail unless and until the organization can modify what it needs to about the culture to support such a strategy.

Now that we have covered why culture matters in change, let's discuss how to determine if the culture within the organization is one of change and, if not, how to get it to be so.

Is the Culture a Culture of Change?

An organization that has a culture of change is one in which *everyone* embraces and welcomes change. Change is seen as a norm in such organizations. This does not come naturally. No organization is, by nature, one that has a culture of change. Rather it is a process that starts at the top of the organization and trickles down to every employee. The most successful, most competitive, and most profitable organizations are ones that have a culture of change. This is different than whether or not the culture is aligned to the desired change. Even in a culture of change, the change initiated *must be aligned* to the culture of the organization or it cannot be successful. Being a culture of change, however, may make aligning the culture a bit easier of a task since the organization is more positive about change overall.

Vincent, a recent graduate of an ivy-league business school, launched his fitness-focused organization early after graduation. He was very much focused on ensuring a culture that matched his vision for the organization. This vision included an organization where employees were empowered to make decisions, solve problems, and take risks to enable continued innovation. He wanted individuals who were collaborative with a shared sense of responsibility and commitment to the organization. Vincent also wanted individuals who were continuously striving to improve how the organization achieved the goals—basically individuals who embraced regular change. Vincent was, from the start, creating an organization that had a culture of change.

Are People Ready for Change?

It is essential to determine if the employees of the organization are ready for change at the time that the change is being launched. For example, an organization that has been through difficult times, such as a layoff due to economic difficulty, is unlikely to be ready for another change initiative soon after the layoff. The layoff was a change in and of itself. In determining readiness for change, organizational leaders should consider the following:

- Will employees accept the change at this time?
- What are the employees' beliefs and attitudes around this particular change?
- What are employees' expectations around this change and are those expectations aligned with leadership expectations?

In addition to what has been shared regarding the change—acceptance and perceptions of change, as discussed in Chapters 2 and 3—consider other options to gauge readiness for change, as shown in Table 4.3.

The more complex and therefore impactful the change initiative launched is, the higher the necessity of determining if the employees are ready for change. If employees are not ready for change, time must be invested in getting them ready, in order for the change to be successful.

Using Information from Surveys and Employee Engagement Data

Let's explore the use of general surveys as well as the use of information gathered from employee engagement surveys. These tools help with understanding whether there is an appetite and readiness for change.

Table 4.3   Determining readiness for change

Formal Channels to Gauge Readiness	Informal Channels to Gauge Readiness
• Employee engagement surveys • Focus groups • Departmental meetings • All-staff meetings	• Informal conversations with employees • Seeking out and having conversations with influential non-leaders in the organization • Gauging the mood of employees through observation

By utilizing data gathered from employee engagement surveys in particular, organizations can learn much about the mood of employees and can also understand where employees believe the organization needs to improve. Here's an example of using employee engagement survey data to determine change readiness and tolerance:

> *A regional medical center sent out annual employee engagement surveys to their 1,200 employees. This practice has been in place for over four years. The survey was sent out three months prior to the senior leadership team gathering for their annual strategic planning session. A subset of the leadership team consisting of representatives from each of the functions evaluated the results of the employee engagement survey and summarized the information for the senior leadership team. In particular, a focus was put on two areas: first, where employees felt improvements were necessary to enable them to more effectively accomplish the goals of the organization; and second, what employees needed from their management that they were not presently getting that would enable them to be successful in their roles.*

This employee engagement survey data provided leaders with areas ripe for change that were necessary to enable the organization—and its employees—to continue to prosper and grow in a competitive marketplace. Figure 4.6 provides a list of the questions asked in this client organization's employee engagement survey that supplied the beneficial data needed to structure change initiatives on an annual basis.

The questions the client asks (in addition to the usual employee engagement questions) are focused on identifying areas where change is necessary to ensure the success of the individual employees and the organization overall. Year after year the organization reviews the responses to these questions to identify patterns where a focus is necessary for increased efficiencies and effectiveness in meeting organizational goals. Responses to

Process-focused
• What processes are inhibiting, or making it difficult, for you to accomplish your work? • Where have you already made some minor improvements in processes you use regularly?
Leadership focused
• How can leadership enable you to more effectively perform your role? • Where is leadership most effective now in enabling you in your role?
Department-focused
• If you could change one thing in how the work gets done in your department, what would you change? • What challenges are faced by your department?

Figure 4.6   Questions on client employee engagement survey (improvement focus)

these questions also provide insight to leadership about the organization that they may not otherwise have access to—given, frankly, their detachment from the day-to-day work of the employees in meeting client needs.

At this client, the leadership believed that since the employees were the ones actually doing the hands-on work of accomplishing the objectives set by leadership, allowing them to outline the change based on their responses to key questions ensured readiness for change. Change was not being thrust at them; rather, they were defining where change was required. This engaged employees in change very early on.

Let's continue our story with a focus on one particular change initiative launched, based on the results of an employee engagement survey:

> One year, after reviewing the data from the employee engagement survey and considering where the organization wanted to focus their efforts for the upcoming year, leadership launched a change initiative focused on evaluating and refining processes and procedures around how new patients were acclimated to the practice. The individuals selected to serve on the project team for this initiative were individuals who represented a number of functions and worked closely with new patients. A positive for the organization was that many of the individuals involved in acclimating new patients had responded in the engagement survey that this was an area that they felt needed attention. However, leadership realized this did not mean that every individual in this area wanted change to happen.

Because the organization in this example utilized information from an employee engagement survey to structure the change, they were, effectively, initiating change that employees wanted to see happen. The number of individuals who did not support the effort—and there were some, of course—was lower than it might have been had the organization simply launched a change initiative that employees did not see the need for. In this case, employees had a vision for improvement in a particular area and leadership used that interest to launch a change initiative that was also in alignment with a need for organizational improvement from their own perspective.

Of course this does not imply that change desired at the employee level will be in alignment with a vision desired at the leadership level, but when this alignment can occur it should be taken advantage of. When the data from employee engagement surveys are tied to changes launched within the organization, the focus will be on a culture of change. Employees are, effectively, dictating much of the change in the organization when leaders can tie what employees have shared in the surveys to launched change initiatives.

General surveys are another tool that can be used to structure change initiatives and to get employees aligned behind necessary change. These may be used to specifically ask the employees questions focused on developing a *strengths, weaknesses, opportunities, and threats (SWOT) analysis* for identifying opportunities for change or to enable employees to rank and help prioritize areas already identified by leadership as ripe for improvement.

For example, let's assume that an organization wants to look at areas of the business where improvements will help to increase efficiencies. There are three areas in particular that, according to the executives, will benefit from increased efficiencies. The executives determined that by asking employees their opinions, they will be able to better prioritize those areas of improvements that will have the greatest positive impact on employees. Figure 4.7 provides a list of potential questions, the results of which the organization might use in order to determine a prioritization of improvement in three particular areas of the business.

Asking for employees to participate in prioritization of key change initiatives allows the organization to not only engage the employees in change, but also to understand where change will have the most positive impact and assist the employees in performing their responsibilities in meeting organizational long-term goals. Responses to the additional questions included in Figure 4.7 will enable the organization to consider other

Leadership has identified three areas within the business that would benefit from process improvement in order to increase efficiencies and effectiveness. These are:
- Sales and marketing and in particular the use of a CRM to improve tracking and relationships with clients. (CRM)
- Refinements to processes in use by customer service reps to enable for improved support in solving client problems. (CUST SERV PROCESS)
- Refinements in the processes utilized to determine product development opportunities to improve the time to market of new products. (NEW PRODUCT ID PROCESS)

Please respond to the following questions based on your knowledge of the business as well as your role and responsibilities within the business.

1. Please prioritize these three potential areas for improvement based on the following factors:
   a. Value to customer
   b. Ability to better perform your role
   c. Addressing current challenges in business that you see on a day-to-day basis

Use a scale of 1 (top priority) to 3 (last priority)

__ CRM

__ CUST SERV PROCESS

__ NEW PRODUCT ID PROCESS

2. Please share more details as to how you prioritized these initiatives.

3. If you believe there are other areas that need to be examined for process improvement initiatives, please list them here and provide a justification for improvement in this area. Specifically consider the positive impact for yourself in performing your role, your peers and/or the organization as a whole.

These additional questions will enable leadership to understand more about perspectives of the business in order to ensure that initiatives launched enable for ensuring the long-term viability of the organization and for supporting you in accomplishing your goals.

4. Given your knowledge of our industry and the marketplace, where do you believe the greatest opportunities lie for the organization? Please be specific in your responses.

5. What challenges do you see the organization facing in the next 3–5 years, and why?

Figure 4.7 Questions on survey to determine prioritization of areas identified for improvement

necessary changes that will contribute to the long-term viability and success of the organization.

Including employees in decision making around change initiatives and in asking for input on what changes might be of value to them and the organization helps to create a culture of change and increases the readiness of employees for change.

The use of general surveys as well as data from employee engagement surveys provides a basis for launching change that has, effectively, been already identified by employees as needed for the organization and for them to be effective in their roles. Change, therefore, does not seem to come out of left field; the employees are engaged in the possibility of change early on and are helping to shape that change.

Creating and Sustaining a Change Management Center of Excellence

CM Centers of Excellence (CoEs) are great opportunities to encourage a culture of change within the organization. A CM CoE is the entity within the organization that is responsible for owning change management and ensuring change management capability. It provides the knowledge around change management through support of change initiatives, regardless of their complexity. In some organizations, this group may be called a Change Management Office (CMO).

One of my clients finally invested in creating and deploying a CM CoE due to the following experiences over a number of years with launching and implementing significant change initiatives:

- *Several high-profile change initiatives that failed at a high cost to the organization*
- *A realization that on a number of occasions various functions were launching change initiatives that were in conflict with change initiatives launched elsewhere in the organization*
- *A lack of attention to engaging employees in change in a consistent manner*
- *Sporadic and ineffective communications around change overall*

CM CoEs provide value to the organization in selecting, launching, and implementing change initiatives in a number of ways, as shown in Table 4.4.

A CM CoE enables accomplishing all that must be done to reach a successful conclusion to a complex organizational change effort. A CM CoE also shows employees that the organization is committed to doing change well by ensuring that there is a framework and structure in place to ensure that change is well managed and well-thought-out.

Table 4.5 provides a number of questions that should be asked in creating a CM CoE.

These questions enable leadership to structure a CM CoE that will meet the needs of the organization in launching and implementing change initiatives, whether over the long term (as a permanent function within the organization) or for management and implementation of one large, complex, and transformational change. Certainly a CM CoE created for

Table 4.4   The value of a CM CoE

• Providing a team of subject matter experts from throughout the organization • Determining change initiatives to be launched through ensuring alignment to strategic objectives • Engaging employees in the change • Solving problems that arise due to the change • Developing a plan to manage the change and ensuring a strategic project management approach • Providing a knowledge repository for all change initiatives	• Providing a central point for all communications regarding the change initiative • Ensuring development of best practices, processes, and procedures related to the change initiative • Ensuring alignment of change to the organization culture • Providing training for those impacted by the change initiative • Providing coaching to leadership and others

Table 4.5   Questions to answer prior to creating a CM CoE

Questions to be answered...	
1	What is the objective of the CM CoE? Is it permanent or solely for the purpose of implementing a specific, complex, transformational change initiative?
2	Where in the organization will the CM CoE reside?
3	Will the CM CoE be staffed or will resources be "borrowed" as needed to fulfill demand? Will external resources be utilized in combination with internal resources?
4	How will the CM CoE interact with, or collaborate with, other such Centers (for example, a Project Management CoE or Project Management Office)?
5	Will the CM CoE also be responsible for continuous improvement initiatives within the organization?
6	How will the results of the CM CoE be measured and evaluated?

managing *all* change initiatives will be far more structured than one set up for a temporary change, albeit one that is long-term and complex.

Figure 4.8 provides one potential framework for a CM CoE. The framework for the CM CoE will be determined based on the responses to the questions posed in Table 4.5.

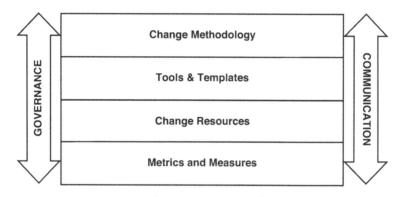

Figure 4.8   A framework for a CM CoE

Table 4.6   A brief description of the CM CoE framework

Governance	• The vision, mission, and objectives of the CoE • How resources (whether budget monies, technology, and/or people) will be allocated to change initiatives • How change initiatives will be selected and prioritized • How lessons learned from change initiatives will be captured and tracked
Communications	• Processes and procedures for communications around change • Engagement of Stakeholder Support Committees
Change Methodology	• Development of standards, processes, procedures, and guidelines for how changes will be managed throughout the organization
Tools and Templates	• Technology, templates, and other documentation to launch, implement, and measure change within the organization • Knowledge repository for change (where tools and templates, guidelines, and other information is retained)
Change Resources	• Roles and responsibilities for change resources • People resources to implement and deploy change initiatives • Management and oversight of Stakeholder Support Committees • Training for those involved in and those impacted by change • Assessments to determine readiness for change
Metrics and Measures	• Determination of metrics and key performance indicators to measure the success of change

Table 4.6 further describes in more detail the framework depicted in Figure 4.8.

Once it is determined that a CM CoE will be of value to the organization, a charter should be developed. A charter formally authorizes the existence of a CM CoE within the organization. It defines the CM CoE in detail and provides authority for the CoE to interact and function within the organization. Table 4.7 provides more details on components of the CM CoE Charter.

Figures 4.9 and 4.10 provide an example CM CoE Charter from a global pharmaceutical organization.

Figure 4.9 provides the first half of the CM CoE Charter for the global pharma company. As can be seen in this figure, the Director of the CM CoE reports directly up to the SVP of Research and Development (R&D), which is where, in this organization, most changes originate. Other changes desired may come from other areas, and these are traditionally funneled through the R&D head. The CM CoE has five full-time employees as is shown under the Roles and Responsibilities section. For very complex initiatives within the organization, the Director of the CM CoE would have

Table 4.7   Components of a CM CoE Charter

Component of Charter	Description of Component
Objective and Mandate	• Description of the mission/objective of the CM CoE (why it exists, what problem it is solving, its value to the organization, etc.)
Organizational Context	• Where the CM CoE resides within the organization (reporting structure, interaction with other CoEs, departments or groups within the organization)
Stakeholders	• Who the CM CoE serves within, or external to, the organization (including the value and support the CoE provides to their stakeholders)
Products and/or Services Offered	• What products and/or services the CM CoE will provide to the organization (including who the products and/or services are specifically provided to)
Roles and Responsibilities	• Defined roles and responsibilities within the CM CoE including reporting structures within the CoE
Critical Success Factors	• The critical success factors that will indicate whether the CM CoE is successful in meeting their objectives/mandate
Measures of Success	• How and when success of the products/services and activities of the CM CoE will be measured

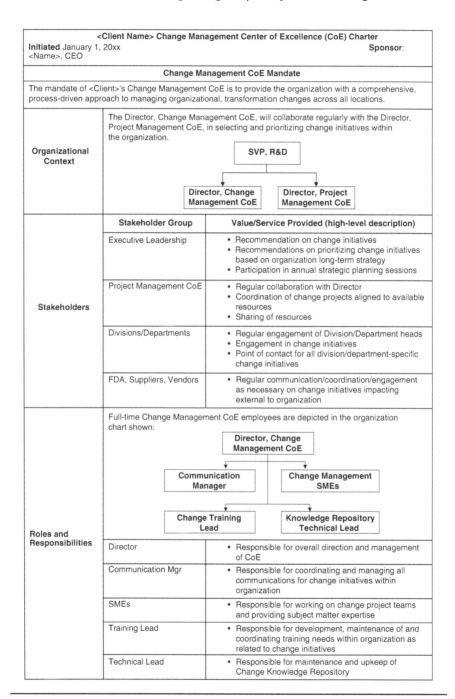

	**<Client Name> Change Management Center of Excellence (CoE) Charter** **Initiated** January 1, 20xx **Sponsor:** <Name>, CEO
	**Change Management CoE Mandate**
	The mandate of <Client>'s Change Management CoE is to provide the organization with a comprehensive, process-driven approach to managing organizational, transformation changes across all locations.
Organizational Context		The Director, Change Management CoE, will collaborate regularly with the Director, Project Management CoE, in selecting and prioritizing change initiatives within the organization. **SVP, R&D** **Director, Change Management CoE** **Director, Project Management CoE**

	Stakeholder Group	Value/Service Provided (high-level description)
Stakeholders	Executive Leadership	• Recommendation on change initiatives • Recommendations on prioritizing change initiatives based on organization long-term strategy • Participation in annual strategic planning sessions
	Project Management CoE	• Regular collaboration with Director • Coordination of change projects aligned to available resources • Sharing of resources
	Divisions/Departments	• Regular engagement of Division/Department heads • Engagement in change initiatives • Point of contact for all division/department-specific change initiatives
	FDA, Suppliers, Vendors	• Regular communication/coordination/engagement as necessary on change initiatives impacting external to organization

Roles and Responsibilities		Full-time Change Management CoE employees are depicted in the organization chart shown: **Director, Change Management CoE** **Communication Manager** **Change Management SMEs** **Change Training Lead** **Knowledge Repository Technical Lead**

	Director	• Responsible for overall direction and management of CoE
	Communication Mgr	• Responsible for coordinating and managing all communications for change initiatives within organization
	SMEs	• Responsible for working on change project teams and providing subject matter expertise
	Training Lead	• Responsible for development, maintenance of and coordinating training needs within organization as related to change initiatives
	Technical Lead	• Responsible for maintenance and upkeep of Change Knowledge Repository

Figure 4.9   Example CM CoE Charter (Part I)

access to other resources (for example, from other departments or functional areas) and may also hire outside resources such as consultants and other experts to assist.

Figure 4.10 continues the charter for the global pharma's CM CoE. Note in Figure 4.10 the critical success factors for the CM CoE. These were particularly important as they would indicate whether or not the CM CoE was meeting its objective and mandate within the organization.

Ensuring Coordination Between the CM CoE and the Project Management Office

It is essential that the CM CoE coordinates closely and is aligned with the Project Management Office (PMO) (or Project Management CoE). In organizations where these two groups or functions are not aligned, it is unlikely that change initiatives will be successful. Remember change initiatives are projects; therefore they must be managed as any other project is managed within the organization. The PMO is often responsible for determining what projects will be undertaken during the year based on the organizational strategy as well as how they will be prioritized, staffed, and scheduled. Change initiatives must be part of this overall project management strategy to ensure that change-focused projects are part of the portfolio of projects to be completed. Figure 4.11 provides one potential organizational structure showing the placement of the CM CoE and its connection to project management within the organization.

As is shown in Figure 4.11, both the CM CoE and the Project Management CoE report up to the Chief Operating Officer. However, there is also an IT Project Management CoE that reports up to the Chief Technology Officer. The CM CoE has a second formal reporting structure—a dotted line to the Chief Technology Officer. This ensures that the CM CoE head is coordinating closely between two distinct Project Management CoEs within the organization.

The CM CoE does not itself manage projects within the organization, but rather should be seen as providing the expertise, resources, and methodology to ensure that change management initiatives are successful. The CM CoE would provide resources to the Project Management CoE for any projects that are change-focused. Let's look at an example:

A global retail organization is launching a large, complex change initiative that will impact a number of functions within the organization. The purpose of the change initiative is to refine processes as well as restructure a number of divisions for increased efficiency.

	Product/Service	Intended Audience
Product/Service	Knowledge Repository, tools, and templates	• All employees
	Training materials	• All employees
	Documented standards, processes, procedures, and guidelines	• Executive team • All division/department heads • Director, Project Management CoE
	Change management personnel	• Executive team • All division/department heads • Director, Project Management CoE
	Change management expertise	• Executive team • All division/department heads • Director, Project Management CoE
	Change strategic plan	• Executive team
Critical Success Factors	The following critical success factors will be used within the Change Management CoE for the first year: • Development of internal change capability • Continuous improvement within organization • Timely and accurate decision-based information about change initiatives • Effective relationships with key stakeholders • Completion of change projects on time and on budget CSFs will be evaluated weekly and updated quarterly for the first year, with updates done bi-annually for years 2 – 4 and on an annual basis after year 4.	

Measures of Success	The following measures of success will be utilized for each product/service outlined in the Product/Service section of this Charter:		
	Product/Service/Activity	**How Measured**	**When Measured**
	Knowledge Repository, tools and templates, training materials	• 100% organization utilization • Ease of use/stakeholder satisfaction • Enables internal change capability • Number of staff trained on change initiatives	Monthly
	Documented standards, processes, procedures, and guidelines	• Meets global organizational needs • Enables internal change capability	Quarterly
	Change management personnel	• Support provided on initiatives • Project contribution	Project-by-project basis
	Change strategic plan	• Alignment to organization strategy • Progress toward meeting long-term goals	Bi-annually
	Support of specific change initiatives	• Stakeholder satisfaction/ employee engagement • Project completion – on time and budget • Time to market	Project-by-project basis
	Change management expertise	• Response time to inquiries • Enables internal change capability	Monthly

Figure 4.10   Example CM CoE Charter (Part II)

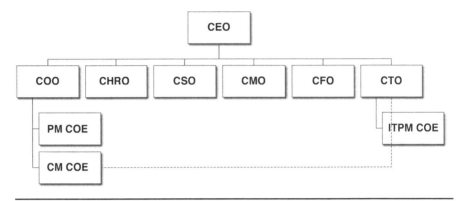

Figure 4.11   Potential organizational structure

> *The Global PMO is leading the initiative; it is, in fact, their larg-est, most complex project for the year. They reach out to the global CMO for support as follows:*
>
> - *Resources with change management expertise*
> - *Assistance in applying the methodology developed to manage change initiatives*
> - *Input into the schedule for the initiative*
> - *Input into the challenges that will be faced within the organi-zation in regards to getting support and commitment for this change*
> - *Development and management of a communication plan to ensure engagement and participation in the change initiative*

As can be seen in this example, the CMO is not leading this project but rather providing support and guidance to the PMO and to the project management manager leading this initiative.

Change is a regular occurrence in today's competitive, global environment in which organizations exist. Organizations that take the time to develop a strategy for building change capability within the organization will create a culture where change is the norm. Such organizations are better able to implement change initiatives that engage large groups of employees and *stick* over time. The change initiatives launched in these organizations are well-thought-out and time is spent communicating about the change before it is officially launched; thereby providing employees the opportunity to help frame the change and contribute to its successful conclusion.

Here's one more story of a major change initiative which a friend and colleague of mine was involved in many years ago while working in the Human Resources Department of a technology company:

Background: In 1988, a Fortune 500 corporation created a world-wide AIDs Program Office as a strategic response to the AIDS pandemic, on behalf of 143,000 employees and their families. With a relatively young workforce and operating in a diverse, inclusive culture, the company was hit with large numbers of men who were out on disability. In addition, there were requests to be transferred, and more dramatic events such as masks and gloves at a metropolitan office and threats of work stoppage at manufacturing plants.

Initially, the company responded by creating "SWAT" teams of employee relations specialists, ombudsmen, and medical personnel to go on-site to quell anxiety. Soon this was neither effective nor manageable. The office was the result of a multifaceted team of stakeholders: employee relations, medical, diversity/inclusion, benefits, the Employee Assistance Program, communications, public relations, law, and senior management. The mission was to develop a coordinated, integrated strategy for responding to the pandemic through enlightened policy, education, counseling, outreach, and philanthropy.

Implementation 1988–1994: The office was staffed with a full-time director (the only one in the country) and a shared assistant.

The work was implemented through management channels. For example, the director would negotiate education programs with line management, starting with the most senior folks and cascading to non-management individual contributors. The education program was a cooperative partnership among local Human Resources, the Business Line, and the AIDs Program Office and rolled out cross-functionally in engineering, manufacturing, and field facilities in the United States. International work came later.

This work was positioned as a business and an employee relations issue, which impacts the bottom line—profitability and health care costs. "We have employees who are ill, or are about to be ill, and we have a business to run. We care. So let's deal with this head-on." (Note: While there was no metric about prevention; it was estimated that by keeping one person from being infected, the company saved an estimated $250K in healthcare costs.)

The education program lasted one half of a business day. There were five major components:

- *Context: Why We Are Paying Attention—always given by the most senior manager*
- *Medical briefing by an HIV/AIDS MD specialist, or reputable MD*
- *Examining psycho-social issues—fear, anxiety, prejudice*
- *Interactive dialogue with the director and two persons living with HIV/AIDS*
- *Next steps: company resources, behaviors, expectations*

(From this list, the dialogue with those dealing with HIV/AIDS had the greatest impact. Initially it was thought that an extensive nursing staff could deliver the medical portion but that didn't work.)

Other activities included family nights and philanthropic investment. For example, the company sponsored annual HIV/AIDS "walks for life" in all major locations where they had a substantial employee base. Over a five-year period, an excess of $5M was allocated in grants, computers, and technical support for major institutions (hospitals, colleges, etc.) and community based agencies.

The program was extraordinarily successful and became part of a transformational culture dealing with illness, diversity/inclusion, prejudice, vulnerability, and individual anxiety. Feedback and evaluations reinforced the value of this intervention. This is not to say that everyone was on board full tilt, but eventually most people bought in through a concerted effort to educate and keep people engaged in dialogue.

Results were as follows:

- *No work stoppage!*
- *Enhanced employee relations*
- *Fewer people out on disability (we encouraged them to work as long as possible)*
- *Appreciation as one of the "best places to work"*
- *New initiatives: company negotiated benefits coverage for HIV/AIDS*
- *Great press; public relations; no missteps re: an explosive issue*
- *Customer consulting; creation of the New England Corporate Consortium for AIDS Education*

Key success factors included:

- *Enlightened leadership re: Corporate Employee Relations*
- *Active support of the CEO*
- *Active support of senior management, modeling the right behaviors*
- *Effective internal marketing—buy in, commitment—"we own this…"*
- *Engagement of key stakeholders from the get-go (medical, employee relations, legal, public and community relations, the Employee Assistance Program, communications, benefits)*
- *And modestly, at a Harvard Business School Conference, a quote from the Corporate Director of HR, "we planned well and communicated well, but we really lucked out by having the right person doing the right work, at the right time. Without that, the whole thing could have blown up!"*

This story provides an example of a tremendously impactful change initiative launched within an organization. In this case, during this time period, some of the readers may recall the fear surrounding HIV and AIDS. A lack of understanding of the disease created fear and, sometimes, irrational behavior that impacted organizations. In this particular organization, the CEO and other senior leaders realized that unless they helped employees across the organization understand HIV and AIDS and educated them about what it meant, there would be a significant impact on their bottom line. Through developing a comprehensive change management program that provided education and understanding across the organization, this particular company was able to reduce the impact that HIV and AIDS could have had on the organization—both on individual employees as well as the bottom line.

This book has free material available for download from the
Web Added Value™ resource center at *www.jrosspub.com*

5

THE VALUE OF FOCUSING ON THE PEOPLE

> "The only way to make sense out of change is to plunge into it, move with it and join the dance."
> Alan W. Watts

DEVELOPING A MINDSET FOR CHANGE

The Chief Executive Officer (CEO) just sent out the announcement to the entire organization that the company has undertaken a large change initiative with a focus on reorganizing divisions and reallocating resources. Sydney knew this was coming. As a mid-level manager responsible for a large team, she had been notified months ago that the leadership team was focusing on an upcoming change that would impact her team. Now the announcement has been made, and Sydney was not looking forward to the upcoming meeting with her team. She knew their tolerance for change was minimal, and this change was a major one!

Chapter 2 discussed the impacts of change on the organization as well as the individual, and Chapter 3 focused on looking at change from a positive perspective. Let's build on enabling change with a focus on developing a mindset around change in the workplace.

The Merriam Webster online dictionary defines "mindset" as "a particular way of thinking: a person's attitude or set of opinions about something; a mental attitude or inclination; a fixed state of mind."

While common knowledge focuses on the definition of mindset as a *fixed state of mind*, it is possible to have a mindset focused on learning and development. Mindset guides what individuals think and how they feel about change. Table 5.1 provides a comparison of an individual with a *fixed state of mind* and another with a *learning- and development-focused state of mind*.

An organization will always have a mix of individuals with a fixed mindset and those with a learning- and development-focused mindset. When engaging employees in change, these mindsets must be considered.

When working within an organization, it should be easy to identify those employees with a fixed mindset as opposed to a learning-focused mindset. Employees with a fixed mindset tend to be more negative around solving problems ("That won't work!") and thus create more conflict due to such negativity. They don't adapt as easily. Those with a learning-focused mindset tend to be more innovative and bigger risk takers ("Let's give it a try!"). They adapt easily and will be excited about trying something new. When changes need to be implemented within the group—for example updating processes—there should be a focus on helping those with a fixed mindset understand how the changes would enable them to reduce their time spent on completing tasks overall; and for those who adapt more easily, the focus should mostly be on enabling them to be more innovative by freeing up some of their time.

Table 5.1   Fixed mindset versus learning and development mindset

Characteristics of a Fixed Mindset	Characteristics of a Learning and Development Mindset
• Hesitant to change • Cannot learn something new • Status quo is best • Too much risk in change • Expectation of challenges and setbacks • Internal voice criticizes	• Change is an opportunity • Desire to learn something new and expand skills • Change is good and provides new skills and knowledge • Risk is acceptable and enables development • Challenges enable growth, setbacks = learning • Internal voice says "you can do it!"

Leaders can enable employees to develop an attitude that is more support-ive of and able to adapt to change through influencing them. Leadership can influence the mindset of employees who are involved in change by being clear about the change initiative and *why* it needs to happen, as well as ensuring an understanding of the vision of the change through framing. *Framing* is a way that leaders can communicate to shape how employees will interpret a change initiative. For example, rather than seeing a problem as a problem, it might be reframed as an *opportunity* and communicated as such. *We have an opportunity to improve how we get work done in the organization* (a positive perception) as opposed to *We have a problem with our processes and they must change* (a negative perception as people may feel they are being blamed or are doing a poor job). Chapter 3 discussed negative versus positive change perspective, which is very much linked to framing communications to achieve a different perspective of change.

> *I have successfully persuaded some clients to move toward a more learning-focused mindset (and be more open to change) by conduct-ing strengths, weaknesses, opportunities, and threats (SWOT) anal-yses within organizations. When I can get employees (and leaders) to see where areas of improvement exist (weaknesses and threats), and where they have strengths and many opportunities to change those weaknesses and move past those threats, I have been success-ful in reframing how they see change for a particular situation. The focus is on the positives and how they can be built on to continue to do better.*

Leaders who are effective at influencing their employees to support change use framing techniques through communicating about a change in ways that enable them to connect with and therefore influence their employees to accept and embrace that change. I know one leader who frames every change in a way that describes it as a win-win situation for the organiza-tion *and* for each individual employee, thereby getting employees engaged in change more rapidly than if he simply discussed the change as a positive for the organization. As another example, I can recall another client who uses sports metaphors to keep his employees moving forward near the end of complex change initiatives: *"It's crunch time! We are on the one yard line with just a few seconds to go. We are champions! Let's do this!"*

Engaging in Conversations Around Change

> *As part of a strategy to regularly engage employees in change, one nonprofit organization held a brainstorming session that included*

employees and volunteers. Prior to the session, the Executive Director divided the employees into five different teams. She asked each team to come up with a problem that they felt needed to be solved or an opportunity that could be realized that would enable the organization to achieve its mission. The day of the session, each team came in with a problem or opportunity and began brainstorming what changes they would make to solve the problem or realize the opportunity. This small nonprofit already had five ideas of changes to be implemented to realize their mission. Employees and volunteers, many of whom tended to be complacent in their roles, were energized in setting change that they wanted to see based on their own ideas.

Engaging in conversations around change—whether that change is happening right now in the organization or may happen in the future—will help to begin the process of transforming mindsets and making it possible for individuals to see change differently. Leaders can also change mindsets to accept and adapt to change when they model an acceptance of change themselves. Engaging in conversations around change *before* the change is officially launched within the organization will encourage the involvement of a broader group of individuals (including those with a fixed mindset that are not supportive of change). Conversations should focus on the value of the change for the individuals—how necessary skills will be developed and knowledge built or shared; as well as how processes, procedures, and systems will be adapted to support and allow the change to be successful. Conversations around change are of key importance *prior* to launching change when leaders need:

- Diverse insights into the change to shape it
- Knowledge from employees
- Innovation in the change
- To ensure that employees are committed so the change will be successful

Table 5.2 provides a number of conversation starters that facilitate:

- Changing mindsets of employees
- Increasing adaptability to and acceptance of change within the organization
- Looking at change as positive and something to strive for

Table 5.2   Conversation starters

Start conversations around change to increase the comfort level with change...
• I have a dream that we could do...
• An idea for this company that really inspires me is...
• Change can be more positive if we did this...
• Just imagine if we were the best in...
• If we were more proactive within our department and less reactive we would be doing this...
• Trends that may impact us if we don't change include...
• We can beat our competition in the market if we...

> *As part of regular monthly meetings with his entire team, Alexander always poses a question that will explore the need to change. Just prior to his last meeting, he asked the team how the department might better collaborate with another group that had just expanded their head count by 50 new hires. He reminded the team that as the other group grew in headcount, it would impact the informal way his team had worked with them in the past. By asking this question, Alexander was pushing his team to look at their current in-place processes and to refine them to continue working effectively with their peers.*

Leaders who acknowledge and appreciate the variety of mindsets that comprise their organizational culture are more likely to understand how to engage those with a *fixed mindset* to enable them to support and adapt to change. Individuals who are resilient (as discussed in Chapter 3), are more likely to embrace, support, and adapt to change. Resilient individuals are those who do not have a fixed mindset. Rather, they realize that by embracing change, they can help to shape it in a way that works for them. They see change as a way to continuously improve and keep moving forward—a learning opportunity. These are the individuals with a *learning-focused mindset*. Leaders who are not resilient will have a difficult time leading change and engaging their employees in change. These are leaders who tend to have a fixed mindset and thus, cannot easily, if at all, support change within the organization. Individuals can change their *fixed mindset* through:

- Setting both short- and long-term goals to achieve
- Controlling their negative reactions and taking time to consider a situation—looking at it from a different perspective

- Building a social network within the organization of individuals with whom they can share information and discuss problems
- Consider how they may achieve more efficiencies in their own workload, regularly evaluating how they are doing their own work

Changing a fixed mindset to a mindset that focuses on learning will lead to increased competency and resilience.

For the balance of this chapter, the focus will be on understanding more about the people of the organization through stakeholder analysis and through regular interactions with people, then using that information to engage them in change initiatives. When leadership approaches change from the perspective of the people of the organization, and framing change in a way that engages everyone from those who support the change down to those who are actively working against the change, they will increase their chances of the change sticking over the long term.

Making It Easier for People to Adopt the Change

Individuals possess eight primary needs that must be met (Maslow's original Hierarchy of Five Human Needs, 1943; Maslow's three additional needs: cognitive, aesthetic, and transcendence were added to the model in the 1960s and 1970s). The priority of these needs is different for each individual; what is important to one individual may not be as important to another. Complex change initiatives trigger a focus on one or more of these needs. If these needs do not appear to be met by the change, then that individual will resist that change. Table 5.3 depicts the final eight human needs defined by Maslow and provides a brief description of each.

Understanding these needs is important in understanding how to motivate individuals to move toward accepting and adopting change within the organization. From a leadership perspective, understanding that these needs must be met in any change and ensuring they are met will enable people to support and therefore adopt the change more readily. Much more can be learned about Maslow's hierarchy of needs through a simple Internet search; however, from a leadership perspective, a base knowledge of these core human needs will enable a better understanding of how to engage individuals in change.

When leaders focus on the people impacted by and involved in the change, they can increase the adoption rate of those who accept and embrace the change. This, however, does not happen without a concerted effort to plan for and engage the people in change. Figure 5.1 provides a worksheet that may be used to ensure an understanding of the change and its impact to assist in preparation of what steps to take to get employees to adopt the change.

Table 5.3 Maslow's hierarchy of needs

Maslow's Hierarchy of Human Needs	
Need	**Definition of the need with a focus on individuals within an organization**
Physiological	The need to survive, for shelter, for a place to belong (this would include the need for a paycheck to survive)
Safety	The need to feel safe, the need for order, rules, and regulations
Love/Belonging	The need for friendship, to feel included and part of a group, to be connected within the workplace
Esteem	The need to feel independent, achievement, and to have control and respect
Self-actualization	The need for self-fulfillment, to reach one's potential, the need for growth and development
Cognitive needs	The need to have knowledge and understanding, to continue to learn and develop intellectually
Aesthetic needs	The need to have balance, an environment that is pleasing and organized; the need to be creative and express oneself
Transcendence	The need to help *others* to achieve fulfillment, reach their potential, and to grow and develop

Documenting this information about the change early on helps with better understanding the complexity of the change and planning for early engagement opportunities. The sooner the organizational leadership can engage employees in change, the better off they will be overall.

One organizational leader knew that she had to modify how changes were launched and implemented within the organization. For the last three years, implementing change was like pulling teeth. Employees fought against every change. In taking a step back and asking her team to evaluate what was going on, the leader learned that, frankly, changes were implemented without thinking about:

- *What else was happening in the organization*
- *What would be done to ensure success of the change*
- *Fully explaining the "why" of the change*

Additionally, and more importantly, by not thinking through the change, leadership was seen as "just launching change to see what sticks." No wonder employees were resisting change!

Through the use of a worksheet depicted in Figure 5.1, the change leader has sufficient information to develop a stakeholder engagement strategy.

Why is the change initiative being launched?	*<Consider: business drivers that are pushing for change, competition, financial stability, etc.>*
What is the vision for the change?	*<Ensure vision supports the strategy of the organization, is clear and understandable, and the organization's employees can achieve that vision.>*
What is the gap between how the organization is today to how we need to be to achieve the vision for the change?	*<Consider all aspects of the business—the larger the gap, the more complex the initiative. Consider also what will be done to close the gap, e.g., training, etc.>*
Who within the organization is impacted by this change?	*<Consider every possible stakeholder—even those with a minor impact.>*
What does success look like?	*<Consider success from the perspective of the organization as well as those individuals impacted by the change project.>*
What details regarding the change can be and must be shared now?	*<Consider process changes, new technology, new opportunities, skill building, etc.>*
What will not change?	*<Focus here on what is NOT going to change, this enables for perspective.>*
How much resistance is expected?	*<Consider how much resistance is expected. The more complex the initiative, the more people impacted—the more resistance that can be expected. Dramatic changes that impact every aspect of how people work within the organization—and potentially whether or not individuals will have jobs—will create more resistance.>*
What else is going on in the organization?	*<Are there other changes in progress? Is the organization in the middle of a "crunch" such as to meet sales goals or get a product out the door? If so, this will impact the ability for employees to engage in this change.>*
What channels exist for communicating about change?	*<Consider a variety of options that might be used to effectively communicate about and engage employees in change.>*

Figure 5.1   Change adoption worksheet

UNDERSTANDING THE PEOPLE: CONDUCTING A STAKEHOLDER ANALYSIS

Every individual with a vested interest in the change, or for whom the change may have an impact, is a stakeholder. For example, if an organization is merging with another and the company name will change based on the merger, a small *to-do* of a much larger change initiative is scripting messaging for the individual at the front desk. The receptionist will be responsible for notifying callers of the new company name and answering questions about the change. This information must be provided to the receptionist so she/

he can do the job. This receptionist is a stakeholder in the change effort. The change manager must get this individual engaged in the change initiative.

Table 5.4 provides some questions to consider in determining whether an individual (or a group) is a stakeholder. If the answer to any question in Table 5.4 is a *yes*, the individual or group should be considered a stakeholder of the initiative.

The bottom line is: every person within or external to the organization who is impacted by the change in some way or needs to provide information for the change—whether that impact on the individual is perceived as minor or not, or the amount of information needed is simple or complex—should be considered a stakeholder and identified as such.

A *stakeholder analysis* is a tool used to understand more about the stakeholders impacted by the change. Diving deep into the stakeholders and their perceptions of the change and how impacted they will be enables the change leader to understand:

- How difficult the change may be to implement
- How much time should be spent upfront preparing for the change *before* engaging employees
- How much time should be spent upfront communicating and having dialogue about the change *before* launching it

Table 5.4   Determining if the individual or group is a stakeholder

Question to consider...
Will the individual (or group) be directly impacted by the change initiative?
Will the individual (or group) be indirectly impacted by the change initiative?
Is the individual (or group) responsible for providing support (training, reports, data), resources, budget money, or any other assistance on the change initiative?
Is the individual a leader in the organization who has the ability to influence the change?
Is the individual an influential non-leader in the organization who has the ability to influence the change?
Is the individual needed to serve on the project team for the change initiative?
Is the individual (or group) an external vendor, supplier, contractor, or consultant who will be impacted by the change initiative?
Is the individual (or group) an external vendor, supplier, contractor, or consultant who needs to provide information or share knowledge regarding the change?

Figure 5.2 is a stakeholder analysis template that may be used to better understand the stakeholders impacted by and involved in a change initiative.

The impact rating of *low, medium,* or *high* is based upon how much impact the change will have on the individual's role and responsibilities or on how they do their job. This is aligned to the amount of efforts required to participate in the change initiative. The more effort the stakeholder needs to put forth for the change initiative, the higher they are impacted by it.

Let's look at an example using a stakeholder analysis template. A national software company is launching a change that will allow some workers to do their jobs remotely. Remote work would be permitted for sales and customer service. Salespeople, rather than coming into the office as they currently do, would have the opportunity to work remotely when they were not visiting customer sites. Office space will no longer be available for them. Customer service personnel would be permitted to work remotely after three years of employment with the organization. Figure 5.3 depicts a partially completed stakeholder analysis based on this example.

As can be seen in Figure 5.3, an initial (partial) analysis has been completed. These perceptions of the change may be gleaned from initial conversations with those impacted or based on leaderships' knowledge of the stakeholders involved.

The information in this stakeholder analysis will be used to develop a detailed communication plan (discussed in Chapter 7). A stakeholder analysis produces an initial understanding of all stakeholders who will be impacted in some way by the change initiative and can be used to determine how early prior to the launch of the change initiative communications and information sharing must take place in order to ensure engagement of the greatest group of stakeholders. Figure 5.4 provides an example of a partially completed stakeholder engagement strategy.

Stakeholder	Impact (Low–Med–High)	How Are They Impacted?	How Much Effort Is Required to Participate in Change Initiative	Perception of Change
<Stakeholder group and/or individual included here>	<How impacted are they by the change?>	<List here the impacts on the individual stakeholder or stakeholder group. Consider what they need to do.>	<Considering what they need to do to participate in the change initiative, how much effort will they have to put forth?>	<Based on the knowledge of the group or individual, what is their perception of the change?>

Figure 5.2   Stakeholder analysis template

Stakeholder	Impact (Low–Med – High)	How Are They Impacted	How Much Effort Is Required in Change Initiative	Perception of Change
Sales Director	High	Develop new sales organization structure Develop new processes and procedures for remote workers Determine use of technology (in collaboration with Technology Group) to enable remote work	*Significant* effort required in design of sales structure as well as working closely with sales to develop new processes and procedures for remote work. Requires keeping salespeople focused on achieving goals through change.	Concerned about losing control over management of sales team if not in office
Salespeople	Medium	Participate in change initiative	*Moderate* effort; while impacted overall, job itself will not dramatically change	Excited about change overall; majority rarely in office now
Customer Service Personnel (Greater than 3 years with company)	Medium	Participate in change initiative	*Moderate* effort; while impacted overall, job itself will not dramatically change	Concerned about not being in office to interact frequently with peers Lack of understanding about need to work remotely
Technology	High	Collaborate with Sales Director to analyze/source/build new technology to enable remote work Train remote workers in use of technology Support remote workforce	*Significant* effort in sourcing or building technology to support remote work; effort impacted by fact that resources are limited in Technology group.	Concern about increased workload and need to manage new technology with limited resources

Figure 5.3   Partially completed example stakeholder analysis

Stakeholder Engagement Strategy		
Change initiative: Remote work for sales and customer service personnel		
Strategy objective: Engage all impacted individuals in the change to ensure a smooth transition from working in the office to working remotely to enable for continued and uninterrupted support of customers.		

Methods/tools to engage stakeholders in change:

- Email
- Internal site
- Focus groups
- One-on-one meetings

- Surveys
- Department meetings
- All staff meetings
- Participation on initiative
- Feedback channels

Pre-change Launch Engagement		
Stakeholder Group/ Individual	**Impact from change**	**Engage In change through...**
Salespeople	High	• Department meetings • One-on-one meetings • Surveys
Technology	High	• Department meetings

Kick off of Change Initiative	
Stakeholder Group/ Individual	**Continue to engage in change through...**
Salespeople	• Internal site • One-on-one meetings • Ongoing meetings • Feedback channels • Email updates • Participation in kick-off meeting
Technology	• Ongoing meetings • Feedback channels • Email updates • Participation in kick-off meeting

Change Initiative Project Work	
Stakeholder Group/Individual	**Continue to engage in change through...**

Launch/Implementation of Change	
Stakeholder Group/Individual	**Continue to engage in change through...**

Follow Up/After Implementation of Change	
Stakeholder Group/Individual	**Continue to engage in change through...**

Figure 5.4   Partially completed example stakeholder engagement strategy

In this example, a determination has been made as to how stakeholders will be engaged prior to the launch of the change initiative and at kick off. A complete stakeholder engagement strategy would follow through from the change initiative project work through implementation of the change and then after implementation. Similar to a communication strategy, a stakeholder engagement strategy should be reviewed regularly and updated as needed. As more information is learned about stakeholders, adjustments will likely need to be made to the stakeholder engagement strategy.

The more complex the change initiative and the more stakeholders involved; the more likely that it is necessary to prioritize the group of stakeholders. This may be done by prioritizing via impact: high impact is a first priority, medium impact a second priority, and low impact a lesser priority. However, the change leader may also then further prioritize stakeholders by the level of power they have within the organization. This may further break down by:

- Priority one: formal power as a leader would have in the organization
- Priority two: informal power (through the ability to influence) that a nonleader may have in the organization

Those with power—whether informal or formal—have the ability to adversely impact a change initiative if they are not kept engaged. Tag them as stakeholders and ensure they are engaged in the change initiative.

> In a survey of ten of my clients regarding the use of stakeholder analysis for complex changes implemented in their organizations over the last four years, clients have noted that they have seen significant improvement in keeping stakeholders involved and engaged in change overall. Client comments from the survey included:
>
> - "We were able to bring along a larger group of employees earlier in the initiative than we were able to do in the past."
> - "After doing an initial stakeholder analysis, we actually learned that we were better off waiting six months before launching the change because there was just too much going on currently and perceptions of change were negative."
> - "A number of employees felt that they had a hand in shaping the change and they perceived we wanted their feedback because we engaged them so early on and, this time, we engaged them in the right way because we thought about how they would perceive the change."

- *"The stakeholder analysis we completed enabled our Internal Communications Group to better plan for communications—including the best way to communicate—about the change early on in the initiative."*

DETERMINING CHAMPIONS, RESISTERS, AND THOSE WHO ARE INDIFFERENT

Employees impacted by change fall into one of three categories:

- *Champions*: those who support the change and are excited about it. These are individuals for whom the change is positive or who see change from an opportunity-driven or positive perspective.
- *Resisters*: those who are fighting against or resisting the change. These are individuals who may be negatively impacted by the change, who are change-weary, or who only see the negative side of change.
- *Indifferent*: those who are unsure about how they feel about the change. These are individuals who may be taking a *let's wait and see* approach to the change or who just don't have enough information to make a decision one way or the other. Indifferent perceptions may also occur in situations where the individual is not impacted by the change at all and therefore is unconcerned about what is going on.

Figure 5.5 depicts a grid showing a suggested framework for engaging stakeholders depending on their current perception/engagement level and the impact the change will have on them.

The higher the impact of the change on the individual, the more important it is to keep the individual engaged in the initiative. However, simply because someone is not impacted does not mean they should be ignored or neglected. An individual stakeholder who is not impacted (low) but is resistant to change may create a situation where he has a negative impact on others.

Consider this situation:

> *Paulina was tasked with reorganizing the sales team. Currently, the salespeople were assigned to regions. Under the new structure, salespeople would be assigned to verticals (finance, retail, manufacturing, etc.). This change was going to have a major impact on the sales team, as it meant that some salespeople would have to shift accounts that they have managed for a number of years over to another salesperson (if the account was not in their newly assigned*

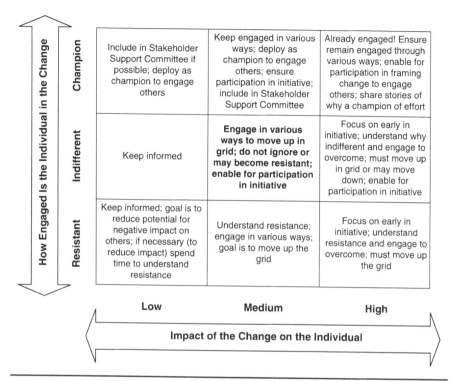

Figure 5.5   Stakeholder engagement/impact grid

vertical). Operations, while their jobs would not be impacted directly, worked closely with sales and, in a handful of cases, may not be working with a salesperson with whom they had a long-standing and strong relationship. Paulina reached out to her counterpart in Operations to discuss what was about to change within sales. She specifically wanted his input on how to keep Operations informed since she knew that some of the operations people would be resistant to this change. Paulina wanted to be sure they were involved in early conversations around the change so they understood why it was happening. The last thing Paulina needed was to have to deal with another group negatively impacting her group when she was trying to get them to accept the change.

When considering who might be a champion, resistant, or just indifferent, consider responses to the question posed in Table 5.5.

Table 5.5 Determining champions, resisters, and indifferent stakeholders

Question to consider...	May be indicative of...
Has the individual actively participated in a positive manner in change in the past?	Champion
Does the individual often initiate change in how he works (continuous improvement)?	Champion
Does the individual seem interested and excited about the change?	Champion
Does the individual understand the vision for the change?	Champion
Does the individual seem unsure or unconcerned, or have many questions?	Indifferent
Is the vision for the change confusing to the individual?	Indifferent or Resister
Are you having a difficult time reading the individual?	Indifferent or Resister
Has the individual been vocal that there is no need to change, things are fine as they are, or that there is too much going on?	Resister
Has the individual worked against past change initiatives, even when the impact on the individual has been minor or there has been no impact?	Resister
Will there be significant changes in how the individual performs his role and responsibilities that will require significant training?	Resister (assumed)

Table 5.5 provides just a sampling of questions to consider when trying to determine how individuals perceive the change. These questions and their answers should, by no means, be considered the only gauge for whether someone supports or doesn't support change. Rather they are just provided as another way to understand who may support the change and who may resist it. All previous chapters have explored ways that would help the change leader understand employees' perceptions of a particular change initiative.

A great way to gauge how many champions a *new* change initiative may have is to consider a *past* change initiative. If any of the following occurred during that past change initiative, the leader can make the assumption that there are more resisters than champions:

- The change initiative was drawn out and there were no quick wins to keep individuals motivated
- There was no sense of urgency from leadership about the change
- After the change was implemented, people went back to doing the work the old way and it didn't seem to matter

- The change initiative was poorly executed and no one understood who was responsible for getting the work done

Leadership that considers past change initiatives, and the impact those initiatives may have had on the individuals of the organization—whether positive or negative—are more likely to focus the next change initiative in a way that is more beneficial (less negatively impactful) for the individual employee.

A recently hired SVP of Operations wanted to make some changes in one area of his responsibility—warehousing—in order to ensure better tracking of product in and out of the warehouses. He spoke with the directors in charge of the warehouses to get their perspective on his goal and to ask for their ideas on how best to proceed. A number of the directors, who had been with the organization for the longest period of time, informed him that such initiatives had been tried in the past and had been less than successful. From their perspective, employees in the warehouses were never engaged in the changes and rather were ordered to follow new processes and procedures. The initiatives were never properly managed, and therefore the new processes were not effective and didn't last. The SVP of Operations used this information from his directors to engage them in assisting him in framing the change to ensure that all of these issues were addressed in the SVP's meeting with his entire staff. At the meeting with the staff, he shared with them that he understood that past similar initiatives had been less than successful and he wanted and needed their support in ensuring the new one would be a success. He was relying on them and their expertise, he noted, to help improve how product was tracked in warehousing. This was a dramatic change for the group. In the past, they were simply told what would change and had no input into the change.

In this example, the SVP of Operations engaged directors early on by sharing his goal and asking their thoughts on the viability of the goal. He used the information he received from them to ensure that he addressed what issues would inevitably arise as issues early on in a discussion with the entire group. By acknowledging the need for their expertise, and asking for their support, he was already setting the stage for a change initiative that would be managed significantly different than the one in the past. He was on track to change a group that would be indifferent at best or resistant at worst to moving toward becoming champions of the effort.

Engaging and Converting (Some) Resisters and Indifferent Individuals to Champions

Let's state right from the start that *not every* resister can be converted into a champion. The goal of the change leader should be to convert as many individuals to champions as possible, but there will always be some resistance. The goal is to manage the resisters to prevent them from bringing others into their group and/or creating too much negativity around the change. This may be done by keeping them informed about the change and providing a variety of channels for them to communicate and share information with the change leader and the change team. Using a variety of channels to communicate (as was shared in Table 3.2) is helpful in moving people from resister or indifferent to champions of the initiative.

> *As part of a merger, a health care organization was closing some facilities and opening a number of others. While no employees in the facilities would lose their jobs, everyone would have to move to other locations. Some would transfer to facilities that were close by, but others would have to move across the state, which would require uprooting their families or leaving their job. Resisters were many! This relocation was just another initiative within a larger change—the merger of the organization with a competitor. Since the beginning, however, the organization neglected to engage the people in the change. Because the merger was a necessity, the assumption of leadership was that there was no need to spend time discussing it—they simply moved it forward, effectively dragging employees along kicking and screaming. To employees, it looked like the relocation was going to work the same way and they had finally had enough!*

When a change is not going well, stepping back and examining what has happened and determining how to fix it and move it forward makes more sense for an organization than to just keep pushing it forward by brute force. Leaders pushing forward with change when there is significant resistance will only serve to increase resistance.

Consider the healthcare organization merger example. If leadership wants this change to go smoothly and reduce the impact on employees' morale and productivity, they need to step back and engage employees in the change. This does not mean to imply that employees can say no to closing offices and relocating; however, they need to be listened to and their concerns addressed.

The more resisters or those who appear indifferent to an initiative, the more time that should be spent up front in sharing the vision for the change and selling employees on the change and why it has to happen. The focus here will always be more on the benefit to employees over the benefit to the organization. Yes, the organization matters, but if leadership cannot discuss the change from the perspective of the benefit to the employees, they won't have champions for the effort. A focus on the people will enable moving people to the role of champions of the change.

If, in discussions around change, it becomes obvious that concerns exist around…

- whether or not individuals have the skills or knowledge to change,
- whether or not the organization has the resources to support the change, or
- whether or not senior leadership is really willing to sponsor and support the change,

…then it is likely there will be a larger number of resisters who must be brought along to support the change.

There is no formula in determining how long to spend discussing a change to engage people in it before starting. There is also no formula that says when the leadership has achieved a specific percentage of people who are committed to change, they have done their job and should launch the change. Table 5.6 provides a number of considerations in determining how long to spend engaging people in the upcoming change to increase the number of individuals who champion the initiative (and thereby reduce the number of resisters).

Let's look at an example of engaging employees in change to increase the likelihood of employees supporting and championing the change. For this client, past change projects did not go well. In particular, leaders often launched change projects without consideration for other work priorities, but rather through trial and error. Basically, changes were launched within

Table 5.6 Considerations in determining how long to spend engaging employees in upcoming change

• The complexity of the change and its impact within the organization	• How much has been changing recently that may cause weariness
• The effectiveness of past change initiatives	• How much employees trust leaders
• Employees understanding of the long-term goals of the organization	• How much is openly shared within the organization
• Is there a common understanding of challenges within the organization	• Are there feedback mechanisms in place and the mechanisms work effectively

this organization to see, as one leader put it, "what will work and what won't." Over time, employees were wary of any changes launched within the organization, as they were not planned well and most failed. Let's look at this client example further:

> The client's last four major change initiatives had not gone well. Employees were tired of change. When one of the directors announced the upcoming change, it was obvious there was going to be resistance. Employees reacted—and not well. One employee summed up the feeling among the group as follows, "Another change that will waste our time and, when it fails—and it will—we'll be the ones at fault!" I met with the client leadership team to review past initiatives and get agreement on what would be done to address issues with past initiatives. Unless and until leadership acknowledged how they created this situation and took steps to resolve it, resistance to changes would continue. Five meetings at various times of the day and during the week were set up to:
>
> - Acknowledge past failed efforts
> - Provide "lessons learned" from each
> - Describe how it would be different going forward
> - Facilitate "complaint conversations" by employees to effectively enable them to speak their mind
>
> Employees were invited to attend one or as many of the sessions as they would like. This time was also used to engage employees in conversations about the upcoming change with a focus on why the change had to occur and to get their thoughts on how to achieve successful change. Our goal was to get as many employees as possible engaged in a conversation around the change and to enable them to see that the leadership had learned lessons from past failed efforts. From here, leadership and I continued to engage employees through small group sessions held on a weekly basis over coffee in the mornings and lunch, as well as through attending department and workgroup meetings. After four weeks of these sessions, leadership launched an internal change project-focused website to continue collaboration and to provide employees a forum for continued communications and input into the change initiative. The change was finally launched three months later with 55% of the employees (champions) supporting the effort. Even after launch, leadership continued to engage employees through weekly meetings over lunch as well as via the internal website. At the end of another three

months, the organization could count 65% of their employees as champions of the effort. Over time, this continued to increase and while the organization never reached 100%, they had launched the most successful change initiative ever.

When an organization focuses on the *people* and gets them engaged in change through sharing information, providing feedback and insight, they are more likely to have a successful change initiative. Additionally, a focus on the people also increases the number of individuals supporting the change initiative overall. Continuing to focus on the people and keeping them engaged enables continuous movement of individuals from *resistant,* to *indifferent,* to eventually, *champions.*

Champions will be identified by their behaviors. These are the individuals who will:

- Motivate others to change by helping them to see the positive side of the change
- Encourage others to participate in shaping and implementing the change
- Have a positive attitude overall
- Regularly look to improve how work gets done in the organization

FINDING YOUR CHANGE AGENTS

Change agent: an individual who enables change to happen by embracing change and helping others to embrace change. Also known as a change champion or a change supporter.

Change agents can be anyone in the organization—a leader, a manager, the project manager, or an individual contributor. In particular, leaders *must* be change agents if they are going to lead an organization in a global marketplace that is becoming increasingly competitive. Additionally, if leadership does not embrace change, change agents—if they exist within the organization—do not stay around for long. Change agents look at future possibilities rather than living in the past and are inspired by what *could be*—for themselves *and* for the organization. As can be imagined, it is difficult to be a change agent in an organization where change is not embraced.

Change agents should come from throughout the organization; don't solely seek out formal leaders when searching for change agents. Yes, they should be change agents and be the face of the change at the leadership level; but recall that in Chapter 1 it was shared that employees are more likely to support and embrace change when they see influential nonleaders support and embrace the change. These are the change agents to seek out—and they are not the formal leaders.

The Role of the Change Agent

The change agent's role during change initiatives is to understand the change—the *why* of the change—and help others to do so. This doesn't mean that change agents love change. It means they come to terms with the change proposed, understand its benefits, and see the value of the change for the organization and its employees. Table 5.7 provides a list of tasks that are commonly performed by change agents within the organization.

As can be seen in Table 5.7, the tasks are focused on:

- Communicating about change
- Listening to others' concerns about change
- Sharing information up, down, and across the organization

Certainly change agents may serve on change project teams, Stakeholder Support Committees, or have other tasks associated with directly implementing the change initiative. Or, change agents may be tasked with spreading the word of an upcoming change throughout the organization and, once implemented, continuing to engage people in the change through casual conversations.

> *Jessie, an executive assistant to the CEO, was having lunch with three of her coworkers. Just the other day, the CEO had announced, during an all-staff meeting, that the organization was purchasing another company. Jessie was excited about this upcoming change, because, as the CEO noted in his meeting, the acquisition enabled*

Table 5.7   Tasks commonly performed by change agents

• Understanding the vision for change and sharing that vision	• Communicating throughout the organization about change
• Understanding the potential impacts of the change initiative on peers and colleagues and working to overcome negative impacts	• Listening to others' fears or concerns about change and helping them to overcome those fears and concerns
• Accepting change and helping others to do so by sharing the reason for change	• Sharing information about the change up, down, and across the organization
	• Championing change to gain support from throughout the organization

the organization to move into a new area of business. Jessie also realized that it meant changes in technology and some business processes to enable better collaboration between the two organizations. However, Jessie's coworkers were less than excited. One of them, a technology analyst for the company, was concerned about what it meant for him as he had no idea what technology would be utilized or even if he could learn what he needed to do. Her other two coworkers supported sales and wondered what it meant for them. Would they have to split up their accounts with salespeople from the other company? Or have to share their clients? They worried this would cut into their commissions. Jessie listened to her coworkers and told them that she understood their concerns. She was excited about it, she said, because she knew it enabled the organization to further grow and compete more effectively. But, she understood that there may be more of an impact on them than on her. Jessie suggested that they each reach out to their managers to get their perspective on the change and the impact to the department. She did caution that it may be that their managers wouldn't know everything yet since the announcement just occurred, however it was always best to ask the questions early on! She noted she was compiling a few questions for her own boss, including asking how her job might change and if she would be supporting other leaders when the acquisition happened. Jessie also reminded her coworkers that there was another meeting at the end of the week to discuss more about the acquisition and its expected impacts as well as to answer any questions. Jessie told her coworkers that she was going to be there, that she had a list of questions, and that she would love it if they would join her with their questions.

In this example, Jessie not only encourages her coworkers to reach out to their managers to learn more but also encourages them to join her at an upcoming meeting to learn more about the acquisition. Jessie is a change agent. She supports the change and explains why she does (growth for the company, ability to better compete in the market) but also notes she has questions too, and would be reaching out to her boss (will her job change, will she support other leaders, etc.).

Table 5.8 shows a number of ways that change agents should be relied upon by the change leader to promote broader employee engagement in the change initiative.

Table 5.8   Ways for change agents to enable acceptance of change

Join conversations	• Engage in informal conversations about the change • Join the conversations already happening around change When peers are discussing the change in the cafeteria, join in the conversation and keep positive about the change.
Speak up!	• Correct inaccurate perceptions about the change • Reach out to senior leadership on behalf of peers when it is obvious that more information is needed When peers say that they won't be able to do the new job because it requires new skills, remind them that training will be provided.
Be a "go to" person	• Listen to concerns of others • Assist them in getting answers to their questions • Be the key resource for peers in understanding the change When peers have questions about change—share what you know, get the answers for them or point them in the right direction.
Be a role model	• Keep a positive outlook about the change so that peers do the same • Don't get involved in negative conversations, rather, point out the positives When peers are negative about the change, share with them the good that can come out of change and help them look at the change from a more positive perspective.
Ask others to help	• Ask peers to help work on tasks related to the change, to serve on a Stakeholder Support Committee or just to provide feedback when asked by management When feedback is requested, ask peers for their thoughts on what is being reviewed/analyzed. Share individual thoughts with them to encourage participation.
Encourage questions	• Ask questions during meetings regarding the change which encourages peers to ask their own questions • Encourage peers to reach out to their management for information During all-staff meetings or focus group sessions, be the one who speaks up and asks questions—even the "dumb" ones—as many others will have the same questions!

Skills, Competencies, and Attributes Necessary to Support Change

Not every individual who supports and champions change should be considered a change agent. Table 5.9 shows some key skills, competencies, and attributes necessary for designating an individual a change agent.

Table 5.9 Skills, competencies, and attributes of successful change agents

• Positive attitude	• Effective communicator
• Able to provide constructive feedback	• High emotional intelligence levels
• Shares knowledge/information willingly	• Seeks understanding and to build
• Informal leader in the organization (*for those not in leadership roles*)	personal knowledge/continuous learner
	• Has built trust with others
• Problem solver	• Ask questions to ensure understanding
• Strong relationship builders	• Reaches out to welcome new hires
• Excellent listening skills	• Critical thinker
• Resolves conflicts with others in a collaborative way	• Team player
	• Presentation skills

As can be seen in Table 5.9, these are the organization's top performers. These are the individuals who can work cross-functionally, pitch in to help when needed, and strive to work toward achieving the goals of the organization. One of my clients sourced his change agents for a change initiative by asking every manager the following two questions:

- "*Within your own group of employees*: If you had to rely on only *one* of your employees, who would it be and why?"
- "*Considering the organization as a whole*: If you could pick two employees from throughout the organization (outside of your own group) to work with you, who would you pick and why?"

From these two questions alone, the client received a list of individuals to serve as change agents. In fact, he noticed that there were many similarities between responses from his managers. The CEO used this information to personally send an invitation to each of the employees to ask them to take on the role of change agent.

Providing a Clear Description of the Role and Training the Change Agents

Just like any other role in the organization, it is essential for leadership to ensure there is a clear description of the role of the change agent and that individuals are trained in performing the role as necessary. Figure 5.6 provides an example of a role description.

This particular role description is focused on a *nonleader* change agent—an individual contributor within the organization. Each of my clients handles the role a bit differently. One client pulls the individual from their

Change Agent Role Description
Overall goal of role: To champion and support change throughout the organization by engaging peers, co-workers, and others in change through a focus on the positive perspective of change for the organization as well as the individuals within the organization.

Report up to: Change Leader

Desired skills and competencies	• Strong communicator (across the organization) • Innovative problem solver • Collaborator (in solving problems, resolving conflicts) • Strong team player • Excellent listening skills • Seeks answers to further understanding • Has built strong relationships across the organization • Has worked cross-functionally (supporting peers' efforts, participating on projects) • The ability to make presentations
Additional attributes	• Willingness to participate in change through serving on change project team and/or on a Stakeholder Support Committee
Expectations in role	• Ensure understanding of the change and why it is happening • Share information between employees and leadership about the change; be the "eyes" and "ears" to ensure leadership understands employee's perspectives, concerns, and fears around change • Participate in formal change meetings • Participate in informal conversations around change • Engage peers and co-workers in change through participating in and initiating discussions in the collaboration portal/internal web site • Seek out answers to pressing questions by peers and co-workers • Push back on leadership as needed! • Provide feedback, support, ideas on the change initiative, necessary training, documentation, etc.

Figure 5.6   Change agent role description

day-to-day job to serve on the team in a change agent role. Another client has individuals serve in the role while still doing their day-to-day work and pays them a bonus for fulfilling the role. And yet another client rotates the role of change agent from year to year with five to eight people designated each year to serve in the role. At this particular organization, the role of a change agent is a great honor and is often the path to career growth and eventual inclusion in leadership programs (for those in the succession plan).

Training for change agents may take many forms and is dependent on the skill level of the change agent. However, if change agents have the skills, competencies, and attributes described earlier, then training is commonly focused on:

- The use of technology to support the change project
- The use of communication tools to be utilized

Continuing to Engage the Champions, Change Agents, and Other Supporters of Change

Once the organization has identified champions, change agents, and other supporters of change, or moved people from being resistant or indifferent to a supporter of change, it is essential to keep them engaged! Consider this situation:

A change leader who was responsible for engaging the stakeholders in the change process was up against a significant portion of the organization that was resisting the change. Two months later, after he was finally able to convert many individual stakeholders to support the change through numerous one-on-one and small group meetings and various communications, he breathed a sigh of relief. He felt he was now able to focus on ensuring the change actually got implemented. He stopped his meetings and various communications and focused on completing the project work. His updates from that point on were primarily focused on reporting the status of the change project to executives. Just three months later, however, he realized that those individuals he thought supported the change now seemed to be resisting it. He just couldn't get participation in implementing the change. He was frustrated.

Certainly the change leader in this situation spent time engaging resisters in the change initiative early on; through a variety of channels. He can't be faulted there. However, once stakeholders were engaged and supported the initiative, he moved his focus away from the stakeholders to the work of the project. He no longer held meetings with stakeholders and reduced communications with them. This only served to push them away from the project once again and revert to resisting the initiative. When this happens, change initiatives are likely to fail. Figure 5.7 depicts a layered approach to engaging stakeholders in change.

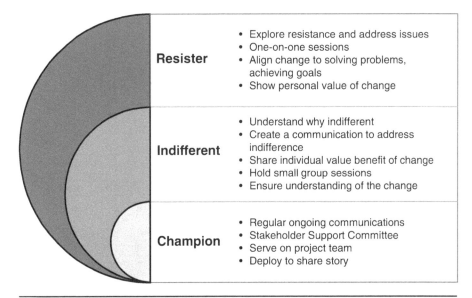

Figure 5.7   A layered approach to engaging stakeholders in change

Starting at the level that needs less engagement (champions) and building upon what will be done to continue to engage them, the change leader will utilize more different and varied frequencies of engagement tools and techniques to move individuals from resister down to champion. Consider this example:

> *Mariah, the head of internal communications, developed a high-level communication plan to engage everyone from champions down to resisters (see Figure 5.8).*
>
> *This plan covered only the first month of communications. The goal was to provide a variety of forums for engaging employees in change. After the initial month, Mariah shared that she would have a much clearer understanding of the support for the change (champions, indifferent, resisters) and could then share with the CEO how to proceed from there. If a number of individuals saw the value and were engaged (champions), she would recommend moving forward with the change during month two. If there were still a number of resisters or those indifferent who could impact the change outcome, Mariah would recommend further engagement before kick off of the change initiative.*

Sender	Audience	Primary Focus	Message Content	Delivery Method	Week (Month 1)
CEO	All employees	Introduce change	• Share vision for change • Why change must happen • How decision was made to move forward with change • Information on follow up meetings for discussions on change (details)	In-person, all staff meeting	First week
Dept Heads	Dept employees	Follow up from all staff	• Share impact of change on department • Benefits to department for change (meet goals, address challenges) • Share benefits to dept employees/potential negative impact if exists (Dept heads were tasked with beginning to categorize employees who were supporters vs those who needed a bit more attention to support the change.)	In-person, department meetings	By end of first week
Head of Internal Comm.	All employees	Launch of internal collaboration/ discussion forum website	• Introduce use of internal site • Promote internal site as a forum for collaboration, share ideas, get answers to questions, participate in change initiative • Provide instructions to access site	Email	By end of first week
Change Leader	All employees	Secure individuals to be part of Stakeholder Support Committee	• Share benefits of Stakeholder Support Committee participation • Share information re: roles and responsibilities of members, time commitment	Email, via department heads, announcement on internal site	Second week
Change Leader, CEO and other leaders	All employees	Enable for further discussions about change project	• **Goal:** sell employees on the change (move resisters and those indifferent to supporting change) • Open group discussions –Q&A sessions	Various small group meetings	Second – Third week
Members of leadership team at all levels, change leader	All employees	Further engage resisters	• Have schedule of various one-on-one meetings to meet with a variety of leaders and/or change leader to have further conversations about change.	Schedule of one-on-one meetings throughout day over a one week period	Fourth week

Figure 5.8   First month: initial communication plan

DRIVING CHANGE FROM THE BOTTOM UP

As was mentioned earlier in this book, change *must be supported* by senior leadership if it is going to be successful. However, senior leadership *alone* cannot drive successful change. Successful change is either accomplished through support of leadership to the lower ranks *or* when driven from the bottom up. Change driven from the bottom up is successful (when leadership will also support that change) because:

- Employees see the need for change and therefore will champion the change
- Employees are framing the change in ways that will work for them
- The change is likely to be completed in a shorter time frame because employees are committed to its success

Driving change from the bottom up, however, does require ensuring sharing and understanding of the vision of the organization from its leaders. Employees who do not know or understand the leadership vision could push for change that may not be aligned to the vision nor the strategic goals of the organization.

Driving change from the bottom up is very difficult, if not impossible, in a hierarchical, top-down, control-based organization. Flatter organizations, where leadership is at all levels and decision making is pushed down to the lowest levels, are better able to support a bottom-up approach to initiating and implementing change within the organization.

One of my clients asked his employees to pick four to six areas of change they would like to see happen. In making their decision, he provided the following parameters:

- *Must be aligned to the vision and strategic plan (which all employees have access to via an internal site)*
- *Must utilize resources already in-house to work on the change*
- *Must remain within the designated budget allocation*
- *Must have support (a champion) at the senior leadership level*
- *Must have cross-functional impact*

This CEO has found that by asking employees to pick areas where they see the need for improvements (change) that would benefit the organization and enable them to be more effective and successful in their own roles, the organization has profited tremendously from those suggested change initiatives that were launched and led by his employees. The leadership team then worked to prioritize the four to *six desired changes proposed by staff, based on certain criteria (in addition to the parameters listed above). These criteria were known to the employees. The list of desired changes was narrowed down to two or three to be accomplished in any given year. Based on the completion of six change initiatives over the last 3½ years, the organization has realized a 10% increase in revenue, a 5% increase in profitability, and cost savings of over $650,000.*

Letting Employees Lead the Change

As can be seen in the client example, letting employees lead the change did not entail letting them move forward with working on whatever initiative they chose or with taking a haphazard approach to approving change initiatives. As a best practice, provide parameters around types of change that would be acceptable as well as well as criteria to enable selecting the best of the change initiatives put forward. Table 5.10 provides some potential criteria for selecting and prioritizing change initiatives put forth by employees.

Table 5.10 is just a sampling of any number of criteria that may be used to evaluate change initiatives that are proposed by employees within the organization. Each criteria should be provided a weight factor and defined globally in order to effectively evaluate the suggested change project against set criteria. Table 5.11 provides an example of that same criteria shown in Table 5.10, but with a weighting factor applied.

Once a change project is submitted by employees, leadership would then use a scale—such as 1–5, where 1 = least aligned to the criteria and 5 = most aligned to the criteria. This scale rating would be used by each individual on the leadership team rating the change project suggestion. Multiplying the weight by the scale provides the prioritization ranking. Figure 5.9 provides an example of using weighting criteria.

Table 5.10   Potential criteria for selecting and prioritizing change initiatives

• Positive impact across multiple functions • Utilizes internal resources • Has cross-functional leadership support • Cost/benefit to the organization • Solves a business problem	• Positive impact to key business processes • Alignment to a strategic goal • Addresses a challenge within the organization • Enables innovation • Allows for taking advantage of an opportunity

Table 5.11   Criteria with a weighting factor and scale

Criteria	Weighting Factor (1–10) 1 = less important to 10 = most important
Positive impact across multiple functions	4
Utilizes internal resources	3
Has cross-functional leadership support	6
Cost/benefit to the organization	8
Solves a business problem	9
Positive impact to key business processes	6
Alignment to a strategic goal	10
Addresses a challenge within the organization	10
Enables innovation	9
Allows for taking advantage of an opportunity	9

Proposed Change Projects		Change Project A	Change Project B
Criteria	Weight		
Positive impact across multiple functions	4	3 (4x3=12)	4 (4x4=16)
Utilizes internal resources	3	2 (3x2=6)	5 (3x5=15)
Has cross-functional leadership support	6	1 (6x1=6)	3 (6x3=18)
Cost/benefit to the organization	8	3 (8x3=24)	2 (8x2=16)
Solves a business problem	9	3 (9x3=27)	5 (9x5=45)
Positive impact to key business processes	6	2 (6x2=12)	4 (6x4=24)
Alignment to a strategic goal	10	4 (10x4=40)	3 (10x3=30)
Addresses a challenge within the organization	10	4 (10x4=40)	4 (10x4=40)
Enables for innovation	9	5 (9x5=45)	2 (9x2=18)
Enables for taking advantage of an opportunity	9	5 (9x5=45)	2 (9x2=18)
	Total points	257	240

Figure 5.9   Example of prioritizing change projects using weighting criteria

As can be seen in Figure 5.9, Change Project A would be the initiative selected based on its total points. Change Project B, however, is not far behind Project A in total points. An organization may choose, therefore, to launch both projects if resources and budget monies allow for it; or a member of the leadership team may work closely with the individual(s) who submitted the initiative to see what adjustments may be made to ensure a higher-ranking project and then resubmit the project for the following year.

Using criteria and weighting factors to select change projects submitted by employees promotes increased fairness in the process, reduces pet projects from moving forward, puts discipline behind the process, and provides a more strategic project management approach to launching change initiatives. Change projects, in this situation, become a part of the overall project portfolio within the organization and are selected and prioritized as any other key initiative would be. Employees feel a part of the bigger picture and are contributing directly to the bottom line when they suggest projects that fit strict criteria.

One of my clients has employees *lead* the change initiative they put forth for approval. Figure 5.10 provides a project organization chart of a

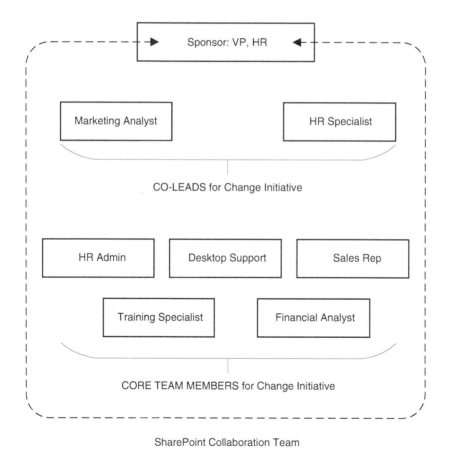

SharePoint Collaboration Team

Figure 5.10   Change project organizational chart

change initiative led by employees with a team comprised of all employees in non-management roles.

As can be seen in Figure 5.10, the team for a Microsoft SharePoint® Collaboration project was co-led by a marketing analyst and a human resource specialist, both of whom worked closely with a team of six comprised of individual contributors from throughout the organization. The team reported up to the Vice President of Human Resources who agreed to sponsor the change initiative when it was brought forth by the employees. The co-leads of the project put together a proposal to launch a new Microsoft SharePoint® site to enable employee collaboration. In particular, employees were interested in a SharePoint® site where they could collaborate through discussions, sharing knowledge and best practices, and problem solving. The change initiative was aligned to one of the organization's strategic goals—increased collaboration across all functions within the organization. The change initiative had significant support among the employees—from individual contributors up to management. However, it would be a significant change within the organization because many of the employees had not used the collaboration tool before and were concerned about expectations of the use of the tool within the organization. In developing the plan for the change initiative, the core team spoke with employees throughout the organization to talk about the value of using the collaboration software tool. They also shared that training would be provided, enabling a broader group of employees to participate in determining how the tool would be used.

Getting Managers Comfortable with Not Leading

For many organizations that attempt to utilize employees to suggest and lead change initiatives, getting managers comfortable is a change initiative in and of itself!

All Company Training wanted to get employees more involved in determining initiatives to be launched within the organization. The employees were certainly excited about the idea—as was obvious from a recently launched employee engagement survey where more than half of the employees selected *ability to recommend and lead change initiatives* as one area in which they would be interested. Management, however, and particularly mid-level managers, were not as excited. Mid-level managers can be significant barriers to change. This is primarily because these individuals are the conduit between senior leadership and individual contributors. If change is moved from being theirs to control and lead, there is a perception of losing control. Therefore, enabling employees to propose and lead change initiatives requires getting managers comfortable with taking a backseat. Putting parameters and criteria in place for selecting and prioritizing change initiatives is one way to assure some level of comfort. Table 5.12 provides some other options to enable management to be comfortable with employees leading change.

As can be seen in Table 5.12, organizational leaders can align managers behind allowing employees to lead change through making *initiating and leading change* components of performance management, employee engagement initiatives, professional development, goal achievement, and/or succession planning initiatives. Let's review how one CEO encouraged his leadership to support change that was initiated and implemented from the bottom up:

> Pablo, a new CEO in a national organization, saw a tremendous need for change. Indeed, in numerous conversations with employees during his visits to various offices, he gathered quite a number of ideas for change from employees throughout the organization. Given the number of change initiatives proposed, many of which he felt would be worthwhile to pursue, he needed employees to drive much of that change. His small leadership team alone could not accomplish what all of those engaged employees would be able to do. However, Pablo knew it wasn't going to be easy to get his senior leadership team to allow employees to initiate and implement change, especially since it meant permitting them to make decisions related to the change. This effort in particular would be difficult

Table 5.12   Enabling management to *let go* and have employees lead change

Suggestions to encourage managers to promote and enable employee-led change initiatives...	
Performance management	Make employee-led change initiatives a component of performance management.
Employee engagement	Include employee-led change initiatives as a component of employee engagement initiatives.
Professional development	Use employee-led change initiatives as a way to enable professional and personal growth opportunities for employees.
Goal achievement	Tie employee-led change initiatives to goal achievement within departments, divisions, and the organization as a whole.
Succession planning	Make employee-led change initiatives a component of succession planning programs.

because the organization tended to be very hierarchical with all decisions coming from the top. Pablo decided that the best approach to take was to communicate with his senior leadership about the impact to the bottom line—if it was possible to launch a number of change initiatives at one time. He also tied his desire to enable employees to initiate and implement change to the fact that the organization was in the midst of developing a succession planning program, and this was a great way to determine the capability of employees. As the organization grew internationally, Pablo noted to the senior team that it would be impossible for the leadership team to continue to maintain complete day-to-day control over every change that needed to be accomplished. Rather, the leadership team would need to be responsible for sponsorship of change initiatives while allowing the implementation of change to be led by employees. This would require training employees in:

- *Decision making*
- *Problem solving*
- *Conflict management*
- *Negotiation*
- *Project management*

Once done, the employees would be able to implement change by managing the day-to-day aspects of the change projects and the leaders would be able to focus on continued expansion of the organization.

Driving change from the bottom up—through enabling, encouraging, and supporting employees to…

- come forward with proposals for change, and
- lead and implement that change effort if approved by leadership

…engages employees in the organization and enables them to contribute in a bigger way than they normally would be able to.

To show further the value of allowing the employees to drive the change from the bottom up, let's review a story from another colleague and friend of mine:

Agile transformations with companies that are managed by either traditional-minded individuals or by those who focus their processes on tried-and-true principles are significantly difficult. Traditional methods for how work is conducted—particularly with regard to software development—do have value and are capable of creating a needed product for a planned deployment. Agile methods augment and enhance the traditional ways by taking work from the macro to micro level. For example, a project plan in the traditional sense outlines everything that is known within an organization from the entire life cycle perspective of the product. Agile methods refine and decompose units of work and focus the product team on delivering incrementally.

During agile transformations within traditional organizations, I discovered the best method for change is to be purposely mutinous with regard to educating a team about a new way to conduct work. The team itself becomes the change agent rather than any one individual. A very experienced software developer approached me one day after meeting with a Chief Information Officer (CIO) who had casually mentioned that he expected the software developer to build a mobile application for two different platforms. The developer had grave concerns about the lack of acceptance criteria, given that the CIO had only stated, 'Wow me.' I jumped at the chance to help the developer transform essentially zero requirements into something of substance. The best approach with mobile software development is to be nimble and adaptive while focusing on your customer. I oriented the software development team toward using methods that provide for progressive elaboration of requirements provided by people who will consume the mobile application being provided.

After orienting the team, I witnessed self-organization occurring and focused intent to educate themselves about delivery methods. The CIO continued down his 'wow-me' track with me, focusing all his attention away from meddling with the developers. The fascinating result was the team became the model for delivering a quality product in a nimble fashion and other teams in the organization suddenly began progressively elaborating their own products. Being champions for change and showing results consistently began to transform the enterprise from the bottom up. Today, delivery of new products is occurring rapidly and the value being provided to customers is demonstrated consistently.

Building and Enabling Change Teams

Change teams should not be confused with project teams, although they do work closely together on projects that are focused on change. The individuals serving on the change team are likely to be working alongside project team members, but they have a different purpose on the team. Table 5.13 provides differences between change team members and project team members.

As can be seen in Table 5.13, the primary responsibility of change team members on a change-focused project is to manage stakeholders and, in particular, their adoption of the change initiative. Let's clarify one responsibility in particular. While project team members are responsible for managing changes to the project—such as changes in scope—the change management team members are responsible for developing and implementing a change management plan as it relates to engaging stakeholders in the change and enabling them to adapt to the change. Figure 5.11 provides a potential organizational structure for a change project.

Figure 5.11 is only one possible organizational structure for a change-focused project. As a best practice, including change management team members on a change project, regardless of the complexity of the project, enables better adoption of the change among stakeholders. Additionally, it enables the project team to focus on the project deliverables solely. Such a division of labor increases the success of change projects, as it permits increased focus on the stakeholders without taking away from getting the work of the project completed. Bottom line, change teams should be a part of *every* organizational initiative since every organizational initiative has some component of change.

Table 5.13 Change team versus project team

Roles and Responsibilities of...	
Change Team Members	**Project Team Members**
• Develop and implement change management plans • Engage employees in discussions around change • Serve as change agents or champions for change • Communicate about the project, sharing information from the project team • Assist project team members through engaging stakeholders in helping with testing • Ensure user training meets stakeholders' needs • Provide feedback from stakeholders to the project team	• Provide expertise on the project • Manage changes to the project's scope • Complete individual tasks associated with project deliverables • Work with stakeholders to ensure that business needs are met • Manage stakeholder expectations about project deliverables • Determine project requirements in collaboration with stakeholders • Communicate status of project deliverables

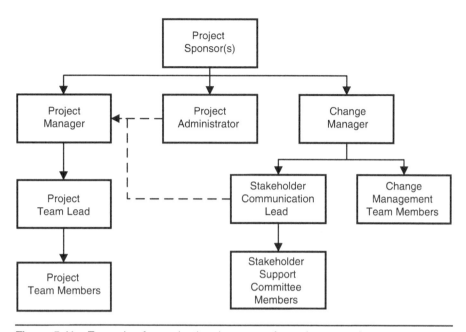

Figure 5.11   Example of organizational structure for a change project

To be effective, the change team needs to have the autonomy to engage employees from throughout the organization in the change initiative, as well as a budget to do so, and support from leadership. It is imperative to have the right people comprising the change team membership. Of course what comprises *the right people* may vary from organization to organization.

Some organizations utilize change teams only for complex change initiatives; while others utilize change teams as a part of *every* change initiative.

Who Should Be Involved on the Change Team?

Change team participations should, ideally, represent every function within the organization. This incorporates a broader stakeholder group and provides for stronger communication about the initiative. This is a great way to spread the word about a change project since each of the individuals will take information about the initiative back to their own departments to share during department meetings or even through informal conversations with their teammates.

One global technology company takes the following approach to change management teams. Around September of each year, the COO sends a message to all employees that the organization is looking to pull together the upcoming year's change management team. This team, he notes, will be responsible for working closely with global project teams to ensure acceptance of initiatives launched within the organization. The change management team's responsibilities will include, but not be limited to:

- *Working closely with the project team to ensure that the change initiative will meet stakeholder needs*
- *Lead stakeholder meetings to gather feedback about the process to understand the challenges and concerns of stakeholders*
- *Be the bridge between the project team and the stakeholder*

Participants on the change management team would be selected from throughout the organization, ensuring representation from all functional areas and all geographic locations. The ideal candidate to participate on the team would:

- *Have built strong working relationships across the organization*
- *Been a change champion in the past*
- *Be a strong communicator and an even better listener*
- *Have a desire to participate on the team with a focus on continuous improvement through change*

The typical change management team includes 12–15 members from the individual contributor ranks within the organization.

In this story, having individuals who champion change as well as individuals who have built strong working relationships within the organization was of primary importance. Participants on change teams are rarely able to tell stakeholders what to do or how to do it; therefore they must be able to influence stakeholders to adopt the change. Influencing requires trust. In order to build trust with others, it is essential to have built a working relationship with them.

While for this particular client it was important to have change champions as part of the change team, another organization may prefer a mix of those who support change and those who do not. Here's an example:

An executive vice president in a financial institution was sponsoring a large change initiative that would impact all functions across the organization. He knew that there were a number of employees who were not happy with the upcoming change. Rather than ignoring them and pushing forward, he wanted to get them involved.

Past experience told him that many of those individuals who were against the change likely had good reasons for why they were fighting the change. By getting them involved, he was more likely to get at what the resisters were concerned about and therefore could address those concerns. Additionally, by enabling resisters to get involved rather than pushing them away—allowing them to voice their concerns and get them addressed—he would eventually move them to champions of the change. This would increase the potential of a successful change initiative.

Wait! Why Have Resisters on the Change Team?

While some change experts will argue that change teams should be comprised purely of individuals who support change; others will argue that having a few resisters on the change team will *only* serve to benefit the team overall. I lean toward the latter. Resisters will *only* be converted to champions when they are able to air their concerns and those concerns are addressed. Smart leaders don't assume that resisters will just go away or eventually turn into supporters of the change through magic. Adding resisters to the change team will generate a number of benefits as shown in Table 5.14.

It is not as simple as just telling a resister that she will be serving on the change team; this will only create unnecessary conflict on the team. Rather, it will be important for the change sponsor to *sell* the value of participating on the change team to the resister.

Each time one of my clients launches a complex change effort, I work with the change sponsor to develop the change team. This particular client looks for a balance of resisters and champions of the change

Table 5.14   Benefits of adding resisters to change teams

Individual (Resister) Benefits	Organizational Benefits
• Ability to get involved in shaping the change	• Enables a broader perspective of the change
• Ability to share concerns and worries around change	• Enables understanding the negative perception of the change
• Approaching resistance in a way that enables professionalism as well as high emotional intelligence	• Increases diversity on the team which enables better decision making and innovative ideas
• Being a part of the solution	• Increases ability to engage resisters who are now visible to leadership
• Improves understanding of the change and the value of it	• Increases ability to manage resistance to the change from throughout the organization

on the team. Given the controversy around many of their change initiatives, it is, unfortunately, not difficult to achieve this balance! At the start of the last change initiative that was launched in the organization, I worked with the Assistant Vice President of Operations to develop a communication to be sent to specific employees whom leadership wanted on the change team. The communication focused on the following points that would serve to entice resisters to join the team—the ability to:

- *provide guidance and direction about the change initiative,*
- *shape the change by sharing ideas and contributing to solving potential problems with the change, and*
- *be a part of the team that would define the future of the organization.*

In particular, when leadership knew specifically that an employee would resist, senior leadership approached that employee directly after the communication was sent to personally ask them to participate on the change initiative.

There will always be resistance to change, even in those organizations where change seems to be done well. Engaging those who are resistant by getting them to serve on the change team will help to transform the general mindset around change, as well as build resiliency in the change. It also conveys to employees that their input, feedback, and ideas are valued. It is important to keep resisters interested by getting them involved in the change. This enables, effectively, for managing the conversations between and among those who resist and those who champion the event; as well as managing perceptions of resisters about the organization, its leadership, and the change initiative.

Accountability Is Key

Accountability is essential on the change team—this holds true for leaders, nonleaders, resisters, champions, or simply indifferent team members. Figure 5.12 provides a checklist for accountability.

If leadership can check any of these items in Figure 5.12, there is a lack of accountability on the team that must be addressed. Accountability is key on any initiative, but is especially essential for change initiatives—and is imperative when the change team is a combination of resisters as well as champions. Change initiatives are difficult in and of themselves. Having change team members who are not committed to participating on the change team or being accountable for the work of the change team only adds to the difficulty of implementing the change.

☐	Throwing tasks that are incomplete or of poor quality "over the fence"
☐	Blaming other team members for mistakes or lack of progress
☐	Lack of progress on the team
☐	Lack of collaboration in completing tasks or solving problems
☐	Lack of effective or sufficient communication among team members
☐	Talking against the change project to others with the organization/actively working against the change
☐	An apparent lack of trust between team members and/or between team members and leadership
☐	Missed deadlines; a lack of urgency in getting work done
☐	Missed meetings; a lack of participation at meetings
☐	Not providing feedback from employees up to the project team and the change sponsor
☐	Lack of collaboration with the project team

Figure 5.12   Do these signs of lack of accountability exist on the change team?

Ensuring clear roles and responsibilities of change team members is essential in order to keep team members accountable. Additionally, ensuring that there are clear processes and procedures for how change team members will interact with project team members as well as engage and communicate with stakeholders (employees) also holds individuals accountable for serving on the team.

> One of my clients has job descriptions for individuals who serve on the change management team. Additionally, the client has clear processes and procedures in place for making decisions, collaborating across the organization, communicating on the change project, and providing feedback, as well as solving problems. The client also launches change initiatives by asking the change team as well as the project team to collaborate to determine how to best approach the change so that it is broken down into smaller components that generate regular successes or "quick wins." To get resisters to be more interested in serving on change teams, leadership approved the use of bonus monies for those individuals who participate on a change team and whose participation and communication on the change is done from a positive perspective. This doesn't mean to imply that resistance is unacceptable within the organization; but rather that resistance is done professionally and as a way to move the change in the right direction. The client noted that when employees care deeply about a change—whether or not it should even happen or whether it just needs to have some "tweaks" made to it in order for

it to be acceptable—they are more likely to be engaged in participating. Figure 5.13 provides a job description for the role of a change management team member used by this particular client.

Job Description: Change Management Team Member	
Reports to:	Change project sponsor
Collaborates with:	• Change Project Sponsor • Change Project Team • Stakeholders • Outside vendors • Other Change Management Team members
Tasks:	Working collaboratively with other change management team members, under the direction of the Change Project Sponsor: • Lead all change management activities • Develop change communication strategy and support all change communications • Develop a calendar of formal and informal events to engage employees from throughout the organization in the change • Provide feedback from employees regarding the change to the Change Project Sponsor and Change Project Team • Support development of testing efforts, pilot group efforts, and training needed to ensure a successful change initiative • Identify resistance to change within the organization and develop plans to mitigate resistance • Identify barriers to change and collaborate with Change Project Sponsor to remove barriers • Other tasks as assigned by the Change Project Sponsor to ensure a successful change initiative
Necessary characteristics, skills, and qualifications:	• Strong relationships across the organization • Excellent oral and written communication skills • A history of collaboration across functions • Excellent presentation skills • An understanding of change and the impact of change on individuals and organizations • An appreciation of the need for change • The ability to look at change from a positive, opportunity-driven perspective • High emotional intelligence • Ability to manage conflicts • Strong problem solving skills • A team player/team leader • The ability to utilize active listening skills to engage employees in conversations around change • The ability to influence others to move toward accepting change through sharing stories and a vision for change • The ability to be resilient and adapt to ever-changing environments

Figure 5.13   Example job description: change management team member

6

LEADING CHANGE ACROSS CULTURAL AND GENERATIONAL BOUNDARIES

"The secret of change is to focus all of your energy,
not on fighting the old, but on building the new."
Dan Millman: *Way of the Peaceful
Warrior: A Book That Changes Lives*

THE IMPACT OF CULTURAL DIVERSITY AND DIFFERENT GENERATIONS ON ORGANIZATIONAL CHANGE INITIATIVES

Earlier in this book, the need for diversity in the employees who are tasked with supporting change was discussed. In particular, as shared in Chapter 5, we discussed that the need for diversity in change agents who will have conversations around change throughout the organization will increase the likelihood of success in implementing change. In general, we can define diversity fairly broadly and include: age, gender, sexual orientation, ethnic background, religious affiliation, and work style. In this chapter, however, the focus will be on *cultural diversity* as well as *generational differences*. Both have an impact on change initiatives and therefore should be explored further. Having stated that, this is not a book on cultural diversity or generational differences; but ignoring either of these in undertaking transformational change initiatives only serves to make the path to change significantly more difficult.

Cultural diversity refers to the values, norms, and traditions that impact how an individual thinks, behaves, addresses conflicts, communicates, collaborates, and solves problems in the workplace.

A generation is defined by years of birth and age groups, as well as events that occur during a particular period of time that may define that generation. For example, individuals who are members of Generation X are shaped, in part, by the fact that many of them had parents who both worked outside of the home.

Individuals have a perception of others based on their particular cultural background and affiliation with a generation. That is simply human nature. Unfortunately, that perception means that individuals may communicate in a less-than-effective manner. This would lead to ineffective problem solving, a lack of collaboration, or unnecessary conflict. Table 6.1 provides select key reasons why cultural diversity and generational differences in the workplace may negatively impact change.

Let's look at an example:

> Siobhan is the VP of Operations in a global organization that is part of the entertainment industry. The last change initiative that was launched, which she sponsored, involved a variety of employees from throughout the organization. This included individuals from the east and west coasts of the U.S., as well as individuals

Table 6.1   Key reasons why cultural diversity and different generations negatively impact change

Cultural Diversity	Various Generations
• Incorrect assumptions about a particular culture	• Misperceptions around ideas of change and support for change
• Expectation to conform	• Insufficient communication channels to engage various generations
• Miscommunication and/or misinterpretation due to language barriers	• Expectations of involvement in change based on generational affiliation
• Bias against the unfamiliar	• Misunderstandings of the different generations and how to engage them in change
• Conflicts due to different values	• Belief that individuals in newer generations may not be able to, nor desire to, contribute
• Lack of contribution to the change initiative	

from India, China, Germany, and New Zealand. Additionally, there were young employees just out of college, as well as "Baby Boomers" and those in between. In thinking back on the change project, Siobhan shared these lessons learned:

"We had a number of issues in getting employees to collaborate effectively. Through observation and surveying participants in the change initiative, we learned a number of things that were, frankly, surprising to us. They included:

- *A number of conflicts that seemed unresolvable because 'everyone thinks so differently.'*
- *Misunderstandings about whether or not someone supported change. As one person put it, 'I just assumed that the young generation loved change!'*
- *Insufficient ways to collaborate given the various locations and preferences among the team."*

It was obvious to Siobhan that cultural and generational differences impacted the ability of the group to work together effectively. More time spent in enabling them to get to know each other would, in her opinion, have been of value.

For change initiatives to be successful, especially in organizations where employees comprise any number of generations as well as various cultural backgrounds, it is essential for there to be an appreciation and acceptance of the differences in others. This requires ensuring time is spent enabling team members to get to know each other, not only to understand their differences but also their similarities; and determine how to best use those differences and similarities to promote change that works for all.

Cultural Dimensions and the Impact on Change

An understanding of Geert Hofstede's theory on cultural dimensions will be helpful in improving how change initiatives are led because it focuses on cross-cultural communications. His work in this area was completed while he worked at IBM. Hofstede added two other dimensions later on; however, here the focus will be on the original four dimensions—as those are the dimensions that the author has seen that create the most impact on change initiatives based on her work with a number of clients.

Table 6.2 provides a brief description of the four original dimensions developed by Hofstede, with a focus on the workplace. Hofstede's cultural dimensions theory looks at the effect of culture on the values of its

Table 6.2 Hofstede's four cultural dimensions

Dimension	Brief Description
Power Distance Index (PDI)	PDI refers to the extent that non-leaders in an organization accept and expect that power is distributed unequally. In Low Power Distance cultures, non-leaders expect that there is some involvement in decision making; whether decision making is pushed down to lower levels in the organization or through providing input and feedback prior to decisions being made by leadership. In High Power Distance cultures, non-leaders expect that decision making is made *only* by leadership as that is their role and right as leaders. High Power Distance cultures may be seen in hierarchical organizations, whereas Low Power Distance cultures are seen in organizations that are flatter in their structure.
Individualism versus Collectivism (IDV)	IDV refers to the extent that individuals feel part of a group ("we") or prefer to focus on themselves ("I"). Those who are more individualistic focus on themselves and believe that they accomplish what they do because of their *own* actions. Those who align to a culture of collectivism believe they have achieved all they have due to the support and help of others. Those with a focus on the group work effectively in teams and focus on the good of the team over the individual.
Uncertainty Avoidance Index (UAI)	UAI refers to the extent that individuals tolerate ambiguity or risk within the organization. Individuals with strong uncertainty avoidance do not manage ambiguity nor risk well. Rules, regulations, and laws are strictly followed. Decision making takes longer as significant data is necessary to reduce risks and eliminate ambiguity. Individuals with weak uncertainty avoidance, on the other hand, have higher tolerance levels for risk taking and for managing through ambiguity. (You may recall that in Chapter 3 the topic of resiliency was discussed. Resilient individuals tend to lean toward weak uncertainty avoidance.)
Masculinity vs Femininity (MAS)	In masculine-focused cultures, individuals prefer assertiveness and individual ambitions to achieve goals. Relationship building is less important in such cultures as is a focus on others' feelings. In feminine-focused cultures, on the other hand, individuals focus on building and nurturing relationships. Feelings of the other party are important in interactions.

members (individuals who reside in that culture or as part of that group) and how those values are linked to behavior in the workplace and in personal relationships. Hofstede's theory explains behavior differences between different cultures.

> *When I work with global clients on change initiatives, I take time to learn about the different cultures involved in the change initiative and where they may lean as it relates to Hofstede's work on cultural dimensions. However, I have learned over the years that assumptions cannot be made! Simply because someone comes from*

*a particular region or country does not mean that they are 100% aligned to that culture. Remember that people are individuals. Use the information from Hofstede's Cultural Dimensions Theory as a starting point **only** to understand the individuals working on or impacted by the change initiative to be launched.*

Taking time to understand these dimensions enables better gauging of support for change as well as how to best communicate around change. To learn more about specific countries and their values across these four dimensions, visit Hofstede's website: https://geert-hofstede.com/countries .html. And, for those readers interested in diving deeper into Hofstede's work, consider purchasing the book, *Culture's Consequences: Comparing Values, Behaviors, Institutions and Organizations across Nations, 2nd Edition* (Geert Hofstede, SAGE Publications, 2003).

The individual leading the change initiative must be one who is appreciative and respective of cultural diversity. Figure 6.1 provides a sampling of statements to consider in understanding a change leader's awareness and understanding of cultural diversity.

Using a scale of 1 (rarely) to 4 (often), respond to the statements on your awareness of and ability to engage others across cultural boundaries…	
Scale (1 – 4)	Statement… I am aware and/or understand that….
	people may share the same culture but think very differently.
	there are a number of challenges due to diversity.
	my values may be different, not better, than others.
	there are situations where my cultural background may impact how I interact with someone else.
	cultural differences may impact how people will handle conflict.
	I need to structure my communications based on what works for others based on their cultural identity.
	I have to adapt how I will interact with others to ensure their comfort level.
	the best change management teams are comprised of a diverse group of people with a variety of cultural backgrounds.
	that to be effective as a change leader and champion of change, I must continue to learn about the variety of cultures with whom I interact.

Figure 6.1   Questions to ask to determine awareness of and appreciation for cultural diversity

Based on my experiences managing change initiatives across global organizations, I have found that...

- communication,
- conflict management, and
- problem solving and decision making

...are most impacted by cultural differences on the change team and among diverse stakeholders.

Table 6.3 provides some ideas on how to generate more effective communication, conflict management, problem solving, and decision making while keeping in mind and utilizing the benefits of cultural differences on the change team.

It is essential to keep in mind that cultural differences impact how stakeholders perceive conflicts, problems to be solved, and who should be involved in making decisions.

Early in my career, my work on a transformational change initiative with one global client really highlighted the need to understand perceptions around conflict. The work on the change initiative involved individuals primarily from the midwest United States; but also some from London, Shanghai, and Mumbai. It was expected that there would be conflict around the change initiative given the diverse opinions about the benefits of the proposed change and on how to approach the change; however, what was not considered was the impact that the varying comfort levels around conflict would have on the initiative. Because a number of participants on the core team were uncomfortable with conflict, they shied away from conversations where it was obvious there was conflict. Since these discussions were not well-monitored (in that the facilitator was unaware that there were some individuals not taking part in the conversations), some decisions were made to resolve conflicts that were not the best that could have been made if more people had shared their knowledge and contributed more fully. So, what was the bottom line for the change initiative? The overall end result had more issues than expected and required significant rework—all of which caused some employee turnover and disengagement of those employees who remained.

A statement as simple as, "Does everyone agree?" can cause significant issues if cultural diversity and its impact on communications is not taken into consideration. For some cultures, stating "no," because you do not

Table 6.3 Ideas to generate effective communications, conflict management, and problem solving and decision making

To be more effective here...	Consider these ideas...
Communication	• Enable for communication channels that accommodate quieter team members and stakeholders, as well as those who share more freely • Ensure communications are clear and concise • Have a variety of communication styles that will work for a variety of people • For larger groups of individuals, use smaller workgroups that enable better engagement of diverse members • Monitor nonverbal behaviors such as facial expressions and body movement as these behaviors may be interpreted differently among various cultural groups • Ensure regular "check ins" with individuals to ensure communications are working for them—is the information sufficient, are the channels used appropriate, etc.
Conflict Management	• Send a survey to all stakeholders to understand perceptions of, and comfort with, conflict • Ensure an understanding of what *is* a conflict and what *is not* a conflict • Ensure an understanding that conflicts are inevitable and are acceptable when not made personal (no personal attacks) • Ensuring an understanding of what is an acceptable conflict • Collaborate with the group to develop detailed processes around how conflicts will be managed • Define roles and responsibilities around managing conflict • Promote the value of conflict! For example, conflict enables new ideas and innovative solutions to problems—help stakeholders see conflict as an opportunity, not a negative
Problem Solving and Decision Making	• Learn about stakeholders to understand comfort level with making decisions (e.g., in some cultures, decision making is solely the purview of a leader) • Collaborate with the group to develop detailed processes for how problems will be solved and decisions will be made, include information on how issues are escalated • Define roles and responsibilities around problem solving and decision making • Determine methods for making decisions, e.g., consensus, expert/leader decides, etc. • Enable for channels for discussing minor issues and provide for input to resolve minor issues that arise • Set the stage early that problems *will arise* and decisions will need to be made to increase stakeholders' comfort level when that happens

agree may be considered rude. For others, a "yes," may simply mean "possibly, I'll think about it." An understanding of the variety of individuals involved in the change initiative, as well as taking the time to build relationships through getting to know them, will result in less confusion and miscommunication overall.

> *Any consultant working on change initiatives should ensure that everyone contributes to making a decision in some way. When individuals feel uncomfortable discussing options to resolve an issue as part of a larger team, the consultant needs to gather feedback one-on-one or via a surveying tool to ensure that the diversity of ideas and breadth of knowledge of a diverse team is considered as part of the final decision to be made.*

Figure 6.2 provides a sample roadmap for ensuring participation in gathering information prior to making a decision when the change team is comprised of individuals from a variety of cultural backgrounds, some of whom may be more comfortable than others in participating in problem solving and decision making.

The roadmap in Figure 6.2 was used on a global initiative that included 12 core change team members and 15 other individuals who worked on a variety of tasks on the project or by providing information as requested. These 27 individuals, from five different countries, represented a variety of functions within the organization. Although the initial team was able to get together in one location for the kick off of the change initiative, they met virtually after that first get-together. One issue which arose was focused on how to best proceed with resolving an issue with a process that relied on a specific technology. This was a fairly large issue and would be costly to the organization if a wrong decision was made. To that end, it was desired that data was gathered from all team members regarding the issue and how best to solve it. A variety of channels were used to gather data: virtual meetings, online surveys, and the opportunity to participate in small group meetings. In this way, the change leader would be able to engage *everyone* in participating in solving the issue, in a way that worked for them. For those who were quieter, they were able to participate via an online survey. For those who preferred a smaller group as it would be more comfortable for them, they were able to contribute in those smaller group sessions. Data gathered from these sessions and the surveys were compiled and shared back out to the group. Once all data was gathered, a group of six individuals from across the organization worked together in order to develop a number of solutions to evaluate. Figure 6.2 does not reflect that a number of solutions were developed for this particular complex issue.

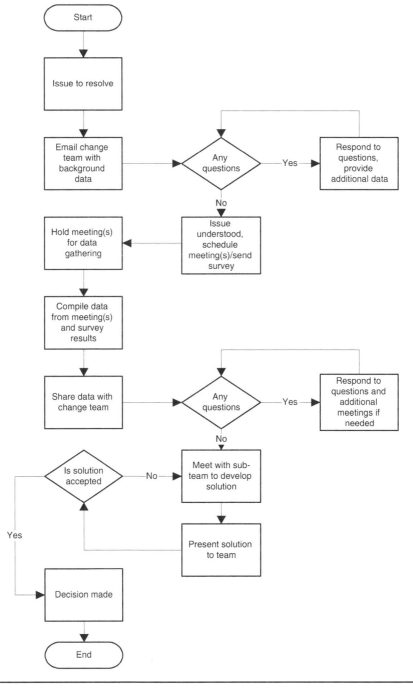

Figure 6.2   Roadmap to ensure participation in decision making

The solutions were weighed against a number of criteria. The core change team members were presented with two potential solutions to the issue and were asked to come to consensus on the better of the two solutions.

By having a process in place *prior* to an issue arising on the change project, issues were better managed, and individuals worked more collaboratively overall. In addition to having a process in place, roles and responsibilities for resolving issues and developing solutions were delineated at the beginning of the change initiative.

Awareness of the cultural differences between members participating in the change team, or those serving on the Stakeholder Support Committee, will promote better engagement of all individuals involved. It also inspires better solutions to problems, more innovative ideas to be brought to fruition, improved collaboration, reduced conflicts, as well as improved communications. Figure 6.3 provides a checklist to use to establish more effective collaboration and communication with diverse change teams.

Generations in the Workplace and Implementing Change

In this section, the focus will be on three generations in the workplace:

- The Baby Boomer generation
- Generation X
- Millennials (or, Generation Y)

	I have learned more about the cultures represented on the change team, but do not make stereotypes based on what I have learned.
	I have taken the time to get to know everyone on the change team as well as enabled the team members to get to know each other.
	I have asked change team members how to best communicate with them.
	I have delineated a variety of channels or methods to communicate with everyone on the team.
	I have emphasized the importance of listening to change team members.
	I prepare for discussions so that I am sure I am providing the right information, at the right time, to solve a challenge or make a decision.
	I understand the value of "putting myself in another's shoes" and request that the change team members do the same.
	I have developed processes for solving problems, resolving conflicts, and making decisions that includes input from a broader, diverse group.

Figure 6.3 Checklist for more effective collaboration cross-culturally

It should be noted that there are still some of the Traditional—or Silent—Generation still in the workplace; however, they are a far fewer number of these individuals given their ages. This generation was born between 1922 and 1945 (although some profiles end the year at 1943 and begin the Baby Boomer generation later). Many of the Traditional Generation members who are still in the workplace are either in senior leadership roles, serving on Boards of Directors, or working part-time. This group values discipline and respects authority. Therefore, when members of this generation lead the organization, there is likely to be more conflict on change initiatives when other generations expect to have more involvement in decision making.

Similar to cultural diversity, the fact that an individual belongs to a specific generation does *not* imply that they are completely aligned with the characteristics of that generation. Every individual is unique and information learned about generations in the workplace should be considered nothing more than a starting point for understanding how to engage that individual in a change initiative. Table 6.4 provides a brief overview of each generation.

Table 6.4   A brief description of the three generations in the workplace

Generation	Brief Description
Baby Boomer	• Born between 1946 and 1964 • Largest generation group • Expect to work hard to achieve goals (often considered "workaholics") • Competitive • High loyalty to organization • Adaptable to change
Generation X	• Born between 1965–1980 • Significantly smaller than the Baby Boomer generation • Seek a work/life balance • Question authority, low loyalty to organization • Value continuous learning • Highly adaptive to change
Millennials (Generation Y)	• Born between 1981–2000 • Very comfortable with technology • Seek a work/life balance • Can be demanding; confident in their abilities • Most educated of the generations • Value diversity and change • Low loyalty to organization

As can be seen in Table 6.4, all three generations are adaptable to change, with the younger generations increasingly more adaptive than the Baby Boomers. Research also shows that all three generations also appear to have similar concerns when it comes to change initiatives in the workplace—such as concerns about change that is disorganized and unplanned or concerns on how to manage through resistance and overcome personal barriers to change. Each of these generations expects and desires two things to accept change:

- Strong leadership driving the change initiative
- Effective and sufficient communications around change

Remember that change is *all about the people*. In particular, when it comes to generations in the workplace, keep in mind that communicating and engaging in a variety of ways is likely to engage the variety of generations in the workplace. Table 6.5 provides a list of some commonly agreed upon core values important to the three generations.

Table 6.5   Core values of importance to the three generations

Generation	Core Values
Baby Boomer	• Involvement/making a difference • Equal rights and equal opportunities • Optimism • Personal and professional growth • Team focused • Work oriented • Success
Generation X	• Appreciation for diversity • Education • Pragmatism • Skepticism • Entrepreneurial, self-reliant • Global outlook • Life/work balance (time value)
Millennials (Generation Y)	• Appreciation for diversity • Multiculturalism • Collaboration/social • Optimism balanced with realism • Confident • Team-focused • Education/continued learning

This is hardly a complete list, nor is what is included in Table 6.5 necessarily a core value of everyone in the particular generational group. When core values are addressed in communicating about the value and benefit of change, leaders are more likely to bring along the various generations to support and participate in that change. Consider this story of a client who needed to engage a variety of generations in a change initiative:

A fairly young Public Relations firm is comprised of employees that break down as follows: Generation X (40%), Baby Boomers (35%), and Millennials (25%). The majority of the executive leadership team, with the exception of the head of finance who is a Baby Boomer, is led by Generation Xers. The Chief Executive Officer (CEO) was in heated discussions with the head of finance regarding how to best move forward with a change initiative. The CEO wanted to establish more effective decision making throughout the organization by pushing down decision making to the lower levels. He felt, rightly, that this would allow for more support from the firm's customer base. He felt the best way to do so was to implement the change via an e-mail to all staff that outlined their areas of responsibility for decision making. The head of finance, however, felt that people needed to be prepared for such a change. While he was not against the change, he didn't believe it wise to move forward so quickly with such a dramatic change in how the organization would, effectively, be run. An e-mail to all employees simply saying that they were now responsible for making decisions to better support customers would raise a significant number of questions and would be chaotic. The head of finance strongly believed that, prior to such a change being launched, the following needed to be accomplished:

- *Alignment of the initiative to the strategic plan*
- *A plan for ensuring that employees had the skills to make decisions*
- *A process for making decisions*
- *An escalation plan for resolving complex issues*

Additionally, the head of finance felt strongly that an e-mail alone was insufficient to engage employees in the change. While he appreciated the CEO's desire to "move quickly to get the change implemented," he also knew from past experiences that much was involved in such a complex change initiative.

If the Public Relations firm wants to launch this change in the right way, they will utilize a number of communication channels to get the message out about the change. In communicating about the change, the executives should include information about:

- Why the change is happening
- The value of the change to the organization *as well as* to the individual employees
- The desire for a cross-generational and cross-cultural team to work on the various components of the change initiative
- Sharing that there will be processes put in place and training provided

Prior to initiating the change, provide a number of forums for Q&A sessions to find out additional information about the why, when, and how of the change.

The variety of generations within the workplace also impacts training that needs to happen to ensure a successful change initiative. Typically, Baby Boomers prefer face-to-face instructor-led training sessions, while Generation X members prefer project-based training with flexibility in learning through a variety of channels, and Millennials prefer technology-based training with experiential activities included. The variety of generations in the workplace should indicate to executives that change projects need to:

- Establish communications through a variety of channels
- Advocate learning to work within the change through a variety of options

Additionally, it is essential that the project sponsor allows the change team to take time early on to understand and structure how they will work together to accomplish goals. In the next section, the focus will be on best practices for engaging and utilizing cross-cultural and cross-generational teams to accomplish complex change initiatives.

While not yet having an impact in the workplace, keep in mind that the next generation is coming soon! The next generation that will impact change initiatives in the workplace in the future is what is being called Generation Z. Generation Z is made up of those individuals who are born after the Millennials. Initial research shows that this generation is fully digital and are the most ethnically diverse group. Given this ethnic diversity, this group will need little engagement and encouragement to understand the value of working cross-culturally. Generation Z members will also be highly educated—even if they do not have formal degrees—due to the ability to participate in free on-line university courses and social learning.

THE VALUE OF USING CROSS-CULTURAL AND CROSS-GENERATIONAL TEAMS TO DRIVE CHANGE

While it may certainly be far easier to push change through with a group of individuals who think and act similarly, it does not guarantee a successful change initiative overall. In this case, a successful change is a change that works for *everyone* in the organization. Cross-cultural and cross-generational change teams will promote a broader perspective on the change initiative. As shared in earlier chapters, a broader perspective stimulates better and more creative solutions.

Figure 6.4 provides a partial list of characteristics that shapes each of the generational groups. These characteristics are focused on how members from each of these generational groups prefer to work, as well as their communication preferences. These challenges—in the differences in preferred communication methods and in the expectations when working with others—will impact the change initiative in a negative way if not addressed prior to launching the initiative.

Baby Boomers
• Desire processes, procedures, guidelines • Prefer structured meetings • Work is serious business, not fun • Prefer face-to-face communications • Detail-oriented • Need to understand big picture and why work matters • Collegial • Formal authority respected • Formal communication channels

Generation X
• Limited processes, procedures • Informal meetings • Want to have fun at work • Use of technology to communicate • Challenge others • Formal authority not desired • Email over face-to-face for communications

Millennials
• Prefer to work closely with those considered friends • Want to understand "what's in it for them" • Want an interactive work environment • Equal to those with formal authority • Email and voicemail over face-to-face • Not great at personal communications due to reliance on technology

Figure 6.4 Characteristics of generations that impact working together

This information enables better planning of cross-functional and cross-generational teams.

> *A 20-year-old technology firm based in the southeast United States was primarily led by Baby Boomers. Over the last few years, however, there had been significant hires from Generation X, mainly due to significant growth for the firm as they expanded their product line and had to replace a number of individuals who were retiring. Additionally, while not a significant size group yet, recent hires were individuals who are part of the Millennial Generation. There were significant clashes between the groups! The CEO noted, "We used to work well together on launching new technology projects, but these days there are so many conflicts! Just the other day I had to break up what I was sure was going to be a fist fight between Bob (a Baby Boomer) and Jeremy (a Gen Xer.) Bob stormed into my office and told me that he has never felt so disrespected by someone and someone better let Jeremy know that he is not high up on the ladder in the organization and needs to start respecting his elders."*

In this situation, even without any change initiatives in place, there were already significant conflicts between generations of employees that needed to be managed. The CEO realized that before he launched the planned change initiative that would focus on evaluating and refining *all* processes focused on getting technology products out to market, he had to resolve this current situation. Let's continue with the CEO's story.

> *The CEO believed it important to enable the various generations within the workplace to spend some time getting to know each other prior to launching any number of change initiatives. This was done through the use of a variety of team building initiatives and other activities that would assist employees in getting to know each other and establishing stronger working relationships and trust. After six months of cross-generational teams working on a variety of small initiatives, the CEO felt confident that he could move toward accomplishing larger change initiatives.*

In particular, cross-generational and cross-cultural teams will make it possible for change initiatives to be better aligned to the needs of the organization—as well as its customers. However, while such diversity adds significant value in increasing innovation, improving knowledge sharing, and improving the overall end results, it needs to be well managed for this to be accomplished. It is simply not possible to take a group of diverse individuals and tell them they are going to work together as a team to accomplish a change initiative and not provide them any support or guidance to be effective in the role.

When leading change initiatives that include team members of diverse backgrounds—as well as team members from different generations—be sure to focus primarily on their similarities, rather than their differences. Too often a focus on the differences makes people feel like they have to adapt to someone else and no one has to adapt to them. While leaders need change project team members to work collaboratively (which requires a bit of give and take), it is essential that the team members understand that they are also similar in a number of ways. Capitalize on those similarities to ensure a strong change project team, while recognizing and appreciating the differences.

Kicking off the Diverse Team to Enable Success on the Change Initiative

Taking the time to ensure a diverse team has what they need to be successful on the change initiative requires planning. There is a big difference between kicking off the project and kicking off the project team. Prior to kicking off the project, the core team members and the extended change project team members must get together to get to know each other and to begin to become comfortable working together. Certainly it is not possible for a team to become completely comfortable working together after just one get-together; but spending time together *prior* to working on the project establishes a baseline to continue to learn about each other and appreciate the differences.

> *I recall a few large change projects with clients where I had to sell to the executives the value of spending time to kick off the project team. For one client in particular, the Chief Financial Officer (CFO) just did not see the need to spend any money on getting the team together in one location to get to know each other. The CFO felt that the team would get to know each other during the project duration. I shared information with the CFO regarding previous change initiatives in which I had been involved. In particular, I was comparing outcomes of change initiatives where the team got together prior to actually starting to work on tasks of the project with projects where they did not meet. Those teams that met prior generally had more successful overall project outcomes, so I encouraged the CFO to give it a try. Fortunately, feedback from the core team (all of whom had worked*

on previous change initiatives in the organization) was very positive about the experience. Comments included, "It was great to actually get to know—on a personal level—the people I would be working with," and, "I never would have guessed about the diversity of our group! We learned information about each other that will make us a stronger team overall!"

Figure 6.5 provides a checklist to ensure that a diverse team starts off on the right path, thereby increasing the success of the change initiative.

Use this checklist prior to scheduling the initial change team meeting to ensure that the time together is effective and accomplishes goals. While it is important to spend significant time on team-building activities and other activities that enable team members to get to know each other, acknowledge similarities, and appreciate and value differences; it is also essential to ensure that actual work associated with getting the project started takes place.

A team member profile is a great start toward building relationships on the team. Providing team members access to profiles *prior* to the first

	Compile documentation available about the project: charter, scope statement, proposed timeline and budget for completion, and other relevant documentation
	Develop a profile for each team member: include a photo, experiences, background, and hobbies (gather information from team members)
	Take the time, as the change project lead, to learn about the variety of cultures and generations represented on the team
	Develop a variety of activities to enable team members to get to know each other during their time together
	Develop an introductory email to send to each team member along with information about their teammates
	Ensure a collaboration tool to enable the team to share information, communicate on the project, and share project documentation
	Develop an invitation list for the initial meeting – including core as well as extended team members, change project leaders, and others
	Delineate what needs to be agreed to on the project, such as processes, procedures, passing work between team members
	Schedule a date and time prior to the change project start date to enable team members to meet each other – face-to-face ideally, virtual if necessary (enable for at least 2 hours together)
	Develop a detailed agenda for the initial team meeting, gathering input from change project leaders/project sponsor

Figure 6.5 Checklist to engage and support a diverse change team

meeting—either via e-mail or through a collaboration portal, will allow team members to begin to get familiar with each other. Figure 6.6 provides a sample team member profile.

One option would be to use team member profiles in conjunction with a "Who Is…" activity to enable individual team members to begin to get to know each other and build relationships. The "Who

TEAM MEMBER PROFILE			
Photo:	<photo>	**Team Member:**	Siraj Hammond
Office Location:	California	**Job Title:**	Director, Engineering, West Coast
Number of Years with Organization:	8	**Department:**	Engineering
Basic Information			
Prior Job:	Worked for a vineyard in Northern California		
Education:	Graduated from the University of California, Los Angeles with a Bachelor's degree in Chemical Engineering		
Most Fun Project:	Working with others in the LA Office to plan our first annual Summer picnic which is now in its 5th year!		
Personal Information			
Birthplace:	Surat, India; but grew up in Leeds in the UK		
Languages Spoken:	English, Urdu, and Punjabi		
Family:	Married with three children, two dogs, and a cat; my wife is an elementary school teacher		
Hobbies:	Spending time with my family hiking, taking walks in the park, and having picnics.		
Best Vacation Ever:	Hawaii! We spent three weeks there and had a great time!		
Fun Fact:	I am a really good bowler and belong to a bowling league.		
What do others assume about you that is incorrect?	I am not a vegetarian!		
Last book you read:	I'm embarrassed to say it, but…I'm reading Hardy Boys Mysteries with my son!		
Skills You Bring to This Initiative/How Can You Help?			
• Project management background • Strong communication skills • Very organized! • Team player			

Figure 6.6   Sample profile of a change project team member

Is..." activity asks questions that require team members to get to know each other by filling out information to respond to the questions. The questions are gathered from a survey sent to all team members asking them to respond to as few or as many questions as they are comfortable with. For example, questions on a "Who Is..." activity could include:

- *Who has a collection of tea cups?*
- *Which three people in the group have gone zip-lining?*
- *Who moved to New York City at the age of 19 with just $300 in his/her pocket?*

A detailed agenda will help to ensure that the limited amount of time spent in an initial team kick-off meeting establishes a balance of team-building activities as well as accomplishing important goals to guarantee that the change project will start off well. Table 6.6 provides a list of items to include on the agenda and/or to provide to change team members during the change team kick-off meeting.

Figures 6.7a and b provide a sample initial team meeting agenda for a change project.

What is not clear in the meeting agenda in Figures 6.7a and b is that the initial introductory activity and the other activities were targeted on understanding the diversity of the team and determining how to best interact and work together relying on that diversity.

Table 6.6   Items to include on the change project team kick-off meeting agenda

• Introductory activity	• A variety of team building activities
• Introduction by Change Leader	• Copies of change project documentation
• Discussion of project goals and objectives	• Discussion around roles and responsibilities for the project
• Identification of potential change project risks	• Discussion around internal and external communications
• The use of Stakeholder Support Committees	• Discussion around how changes will be managed on the project
• Discussion on how the team can support enabling stakeholders to adapt to change	• How information will be shared among change team members
• How work will be passed between team members	• Processes and procedures for team communications, solving problems, resolving conflicts, and making decisions
• The use of a collaboration site and a project repository	• Ground rules for change team members
• A schedule of future project team meetings	• Question and answer (Q&A) session

Project Name:	Project Customer Satisfaction (Cross-Functional Process Improvement Initiative)		
Change Project Leader: Arnold Mellon, SVP, Operations	**Project Manager**: Sally Jackson-Adams	**Date(s) of Meeting**: \<Month/Day/Year\> - \<Month/Day/Year\>	
Meeting Location: NYC Headquarters			
Core Team Members:	• Alison • Becky • Johnny	• Jeremy • Milan • Hamish	
Extended Team Members:	• Sarah • Sam • Gabrielle	• Siraj • Dimitri • Amelie	

Prior to Day One		
Agenda Item	**Timing**	**Ownership**
Dinner and team member introductions	7:00 PM – 9:00 PM	N/A

Day One		
Agenda Item	**Timing**	**Ownership**
Breakfast	8:00 AM – 8:45 AM	N/A
Introductory Activity	9:00 AM – 9:45 AM	Project Manager
Kick off of project	10:00 AM – 10:30 AM	Change Project Leader
Break	10:30 AM – 10:45 AM	N/A
Discussion about project/project documentation	10:45 AM – Noon	Project Manager, Core and Extended Team Members
Lunch	Noon – 1:00 PM	N/A
Continued discussion about project/project documentation; roles and responsibilities on team	1:00 PM – 4:00 PM	Project Manager, Core and Extended Team Members
Break	4:00 PM – 4:15 PM	N/A
Team Building Activity	4:15 PM – 5:30 PM	Project Manager
Dinner	6:30 PM – 8:30 PM	N/A

Figure 6.7a   Sample agenda for a change project initial team meeting

Taking the time to kick off the change project team prior to getting the team started on completing tasks of the project enables the team to increase their comfort level with each other, which means they are more confident working together. This is the perfect opportunity to come to agreement on how the team *will* work together and how they will resolve problems and reach decisions. This time together provides the project manager with

Day Two		
Agenda Item	**Timing**	**Ownership**
Breakfast	7:30 AM – 8:00 AM	N/A
Team Building Activity	8:00 AM – 9:00 AM	Project Manager
Discussions on how team will collaborate and share knowledge	9:00 AM – 11:00 AM	Project Manager, Core and Extended Team Members
Break	11:00 AM – 11:15 AM	N/A
Processes and procedures (includes working lunch)	11:15 AM – 2:15 PM	Project Manager, Core and Extended Team Members
Team Building Activity	2:15 PM – 3:15 PM	Project Manager
Break	3:15 PM – 3:30 PM	N/A
Technology introductory/ learning session	3:30 PM – 4:45 PM	Technology Lead
Meeting Close/Next Steps	4:45 PM – 5:30 PM	Project Manager

Figure 6.7b   Continuation of sample agenda for a change project initial team meeting

a deeper understanding of the team he or she will be leading and often highlights areas where there could be conflicts between team members. Additionally, better decisions are made and less conflicts occur when the change project team has outlined and agreed to—*prior* to the start of the change initiative—how they will manage conflicts, resolve problems, and make decisions. It is much easier to set processes and procedures *before* those processes and procedures are actually needed!

As part of the kick-off meeting, consider holding discussions around not just how the change project will be measured for success, but how the interactions among and between diverse change project team members will be measured for success. These measures may include, for example:

- Fewer conflicts
- Less time needed to solve problems
- Less time needed to complete tasks
- Increased reliance on other team members
- Improved engagement of stakeholders

Over a number of change projects, improvements should be seen over all measures related to improvements in the management of culturally and generational diverse change project teams. Regular evaluation (lessons learned) will result in continuous improvement. Diversity—both in generations and in cultures—is not going away. Those organizations that can embrace these differences and learn how to engage individuals to make the most of the differences will find they are the most successful in implementing transformation change initiatives.

> *One of my clients, who takes a strategic project management approach to all change initiatives within their organization, has a database of information about each employee that enables the change leaders to more easily bring together a diverse group to work on change initiatives. This database allows the client to identify potential team members based on several factors, including, but not limited to: age, gender, time with the company, ethnic background, types of past initiatives in which the person has been involved, special skills, educational background, and roles and responsibilities within the organization. This same client has found that initial team kick-off meetings have shown the following benefits: reduced conflicts on the team, increased collaboration on the team, and more effective problem solving.*

This book has free material available for download from the
Web Added Value™ resource center at *www.jrosspub.com*

7

THE CHANGE PROJECT

"They always say time changes things, but you actually have to change them yourself."
Andy Warhol, *The Philosophy of Andy Warhol*

THE BASICS OF PLANNING ORGANIZATIONAL CHANGE INITIATIVES

I have worked with organizations that manage change well and others that don't. In looking at large global clients over the last five years and their effectiveness at managing change, those organizations that do change well have the following in place:

- *A strategy around managing change that builds on their project management strategy*
- *A project manager to manage the project (this includes the tasks, budget, resources, time, and scope of the project)*
- *A change manager to manage the people-side of change (this includes engaging and encouraging people to support and adopt the change)*
- *A sponsor for the project (project sponsor)*
- *A sponsor for the change components (change sponsor or leader)*

For those organizations that don't do well managing change, it is because change gets pushed to the side. It becomes secondary to completing the project on time and on budget. However, if the change is not adopted, then the project will not be successful—even if it is on time and on budget. While it is highly unlikely that 100% of stakeholders will immediately embrace change, the more time spent in preparing the people (stakeholders) for change, the higher the number of those who will accept change and embrace it.

Every change initiative must be managed as a project. This is because *all* change within the organization happens through projects. However, managing change projects is a challenge! It is much easier to manage tasks than it is to manage people. And that is what change is about—managing people. People are unpredictable. While there are best practices that can be put in place to better manage people during change, there is no magic formula for doing so. Change management plans that were developed for one change project may have to be altered dramatically for the next change project.

Unfortunately, too often in organizations, changes are launched by the leadership team without considering:

- It is a project that must be managed as any other project
- It requires significant planning up front to ensure a successful end result (acceptance and adoption of the change)
- It requires engaging employees in the change early on and throughout the project, as well as checking in after go-live
- The project needs a change manager who understands that:
  - ◊ Change is a process
  - ◊ Every individual experiences change differently
  - ◊ Change management tasks must be focused on where individuals currently are in their perceptions about the change
  - ◊ People are *the most important component* of ensuring change initiative success

Leadership may not consider that a change must be managed as a project because they know that project management can be predictable and a formula—or processes and procedures—can ensure success from project to project. Throw the word *change* in there, add people, and it's a mess!

Those organizations that do change well, do not separate change initiatives from other projects within the organization. Not only do all changes happen through projects, but every project brings change to the organization. Projects are launched for any number of reasons, including to:

- Improve a process
- Expand services or product offerings
- Merge divisions within an organization
- Respond to customer demands
- Meet a regulatory requirement
- Address new competition in the marketplace
- Address declining market share
- Open a new location
- Update technology

Any of these projects require changes within the organization. They all require people to change how they work or interact within or external to the organization. They also require other potential changes that will impact the people. For example, a project to expand product offerings might require not only developing a new product to offer to the customer, but may also require new technology to be implemented to support that product; training of employees to support that technology to sell the product or provide product support; and/or new manufacturing partnerships to develop the product.

Project teams often consider the end result of the project—what are they trying to accomplish and what exactly needs to be done. Change teams, however, focus instead on the people involved. Rather than looking at the end result—what needs to be accomplished or the goal—they consider how the people have to change for that goal to be achieved. They consider, for example:

- The difference or gap between the *now* and the *future*
- What else is going on that may impact how people perceive the change
- How much must change—tasks to be done, processes, skills, knowledge

All of this is what, in the opinion of change teams, drives *when* a change can be implemented. Effective leaders know that for a change to be successful, a big part of the timeline must be engaging and bringing along the people and not solely the tasks to be done.

Change initiatives require significant engagement early on, during and even after the project itself has terminated—all via a multitude of communication channels. Business process improvement initiatives represent significant change for employees and require a more personal touch. The same is true for any other complex or strategic change initiative—whether business-process-focused, people-focused, customer-focused, or any other type of change within the organization.

The focus of this chapter is *not* to provide the reader with a step-by-step plan on how to manage their projects, but rather to provide the basics and best practices for managing *change* projects in particular with a focus on how to approach projects managing change initiatives from a strategic viewpoint. People and their perceptions of change must be managed if change initiatives are to be successful in the organization. This chapter will build on much of what was discussed previously in this book. Figure 7.1 provides a framework for leading change projects, regardless of the complexity of the initiative.

Figure 7.1 delineates several major deliverables for each of the three key steps in initiating and launching strategic change initiatives that were first introduced in Chapter 2 (Figure 2.7).

> *In ensuring a readiness for change at a pharmaceutical organization moving from research to development, I worked closely with the senior leadership team to frame what the change meant for the organization and its employees. In particular, early communications were focused on sharing the fact that, due to the excellent work done by employees, the organization was moving to the next level and needed to prepare to ensure they were successful in developing drugs. A simple survey sent to all employees helped to gather data that made it possible to identify, early on, who supported the change and who were more concerned about (or might be resisters to) the change.*

Change leader(s) selected to take the lead on driving stakeholders to support and embrace change *must* be individuals who:

- Are respected within the organization
- Are proponents of change
- Can create and share a vision for change
- Can influence others (outside of a formal leadership title)
- Have built strong relationships throughout the organization

Change leaders do not, necessarily, have to be senior leadership team members within the organization; however, they should have a formal leadership role (director or manager level for example) and be known as a champion of change.

The project manager assigned to lead the day-to-day work of the change initiative should also be a proponent of change, as well as have skills in

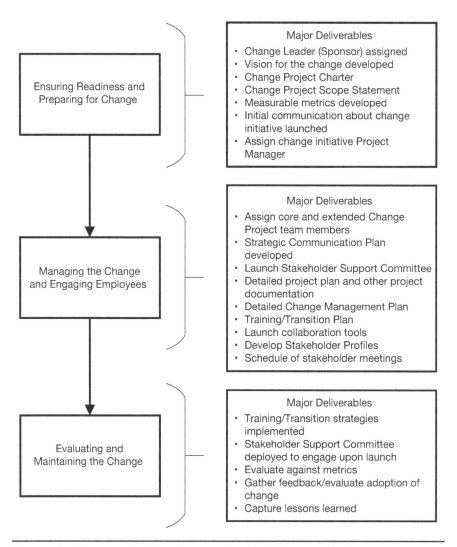

Figure 7.1   A framework for leading organizational change projects

leading projects that are more complex and more prone to risks and failure—as change initiatives often are *if* sufficient time is not spent up front in engaging the organization around the change. The project manager managing a change project should be able, and willing, to work collaboratively with a change manager in order to accomplish the tasks of the initiative (getting the work of the project done) while driving people toward adopting the change.

Outside of the regular project plan and project documentation that should be completed for *any* project launched within the organization, additional components of project documentation that are certainly required for change initiatives include:

- Detailed (people) change management plan
- Training plan
- Pilot testing plan
- Transition plan or post-change implementation plan

Each of these key components of project documentation will be discussed further in this chapter.

While an organization may launch any number of projects in a given year, there needs to be significant thought around the types and number of change-focused projects launched. Figure 7.2 provides a checklist of items to consider to determine the number of change-focused projects that may reasonably be launched within an organization in any given year.

There is no magic number of *yes* responses or *no* responses that indicate definitively whether or not an organization should move forward with launching a change initiative. Rather, this checklist should be used as one point of analysis and research to determine the ability and appetite of the organization to add another change to the list of change initiatives already launched or planned to be launched in the future. From personal experience, I have worked with organizations that can make such a great case for a change to be launched as well as take the time to engage employees to get on board with the proposed change that they seem to be able to manage a significant number of complex change initiatives over a year's time period. In such situations, the organizational leaders ensure that a variety of resources are deployed on these projects and that no one business unit, division, or department is overly impacted by the change initiatives. In other words, they spread change initiatives throughout the organization so everyone feels a part of the progress and no one group feels overwhelmed by ongoing (and likely the appearance of chaotic) change.

On the other hand, I have also engaged with organizations that, as much as they would love to launch yet another change initiative, they simply cannot, or should not, do so. Their lack of processes and best practices around launching and implementing change makes every change initiative—even the smallest, least complex one—a tremendous effort within the organization. In such cases, these organizations can manage at the most two change initiatives a year. And, unfortunately, for a few of these organizations, those two change initiatives they launch can cause great pain among employees.

Proposed Change Project <name here>	
Enter "*YES*" or "*NO*" in the column on the left.	
Change in the Organization	
	Has there been significant change in the organization over the last 6 months to a year?
	Is the proposed change very complex?
	Have the last few change initiatives been successful?
	Were stakeholders sufficiently engaged in previous change initiatives?
	Is there a vision for the proposed change?
	Can the vision be easily shared throughout the organization?
	Is the proposed change aligned to the strategic goals?
	Is there anything else going on in the organization that may impact commitment to and ability to successfully implement this proposed change?
Resources Needed	
	Are there internal resources to manage the proposed initiative?
	Can external resources, if needed, easily be sourced?
	Are resources currently deployed on other initiatives?
	Does senior leadership have the time to devote to the proposed change initiative?
Leadership and Stakeholder Support	
	Do the majority of senior leadership team members support the proposed change initiative?
	Is there broad support for the proposed change among all stakeholder groups?
	Is there time to commit to getting support *prior* to launching the actual proposed change initiative?
Budget Available	
	Is there sufficient budget available to promote, launch, manage and implement the proposed change initiative?

Figure 7.2   Items to consider when determining how many change initiatives to launch

A global training organization had already launched three change initiatives six months into the year. Two of the three had been completed successfully and the third was on track for completion but had a few issues during its implementation. In addition to the change initiatives that had been launched, a number of other projects had been undertaken within the organization. Many of these had been focused on requests from customers. Additionally, the organization had opened three new offices. The Chief Executive Officer (CEO) proposed yet another change to be implemented immediately. The goal of this change initiative was to improve processes with a goal of reducing expenses for training workshops delivered to the general public. Prior to the launch of this initiative, the Chief Human Resources (HR) Officer, given a number of comments she had heard from employees, proposed doing some analysis prior to the launch of the initiative, and worked through the checklist in Figure 7.3. She shared this information with her peers on the executive team.

The bolded "Yes" or "No" responses in Figure 7.3 were concerning to the Chief HR Officer. In particular, she was concerned that resources were already stretched thin and could not easily be deployed on other initiatives. Additionally, given the fatigue within the organization due to the launch of other complex change initiatives, much engagement would need to happen in order to get people onboard with this proposed initiative. And she was unsure even that would help. Given the desired impact from this change by senior leadership, it was obvious that there was limited time to fully commit to getting support. And, certainly not least among her concerns, was that there was no budget for this initiative; although it was aligned to strategic goals. The Chief HR Officer reminded her peers that during the last strategic planning meeting this particular initiative was discussed and, through consensus, put on hold until the following year so that sufficient resources as well as budget monies could be set aside to focus on it. After much discussion, it was decided that the initiative would be put on hold until the start of the new calendar year.

Figure 7.4 provides a more detailed framework for leading change projects (building upon the framework introduced in Figure 7.1).

Figure 7.4 breaks down a change project into two components:

1. Project-specific components and major deliverables in each of the three project phases depicted

Proposed Change Project Process Improvement Initiative	
Enter "YES" or "NO" in the column on the left.	
Change in the Organization	
Y	Has there been significant change in the organization over the last 6 months to a year?
Y	Is the proposed change very complex?
Y*	Have the last few change initiatives been successful?
Y	Were stakeholders sufficiently engaged in previous change initiatives?
Y	Is there a vision for the proposed change?
Y	Can the vision be easily shared throughout the organization?
Y	Is the proposed change aligned to the strategic goals?
Y	Is there anything else going on in the organization that may impact commitment to and ability to successfully implement this proposed change?
Resources Needed	
Y	Are there internal resources to manage the proposed initiative?
N	Can external resources, if needed, easily be sourced?
Y	Are resources currently deployed on other initiatives?
Y	Does senior leadership have the time to devote to the proposed change initiative?
Leadership and Stakeholder Support	
Y	Do the majority of senior leadership team members support the proposed change initiative?
N	Is there broad support for the proposed change among all stakeholder groups?
N	Is there time to commit to getting support *prior* to launching the actual proposed change initiative?
Budget Available	
N	Is there sufficient budget available to promote, launch, manage and implement the proposed change initiative?

*For the most part, some issues on the most recent change, yet to be launched, has caused some fatigue on the change project team

Figure 7.3   Example of completed checklist: global training organization

2. Change-specific components and major deliverables in each of the three change project phases depicted

The primary purpose of Figure 7.4 is to provide the reader with an understanding of the complexity that change projects add to the project management process overall.

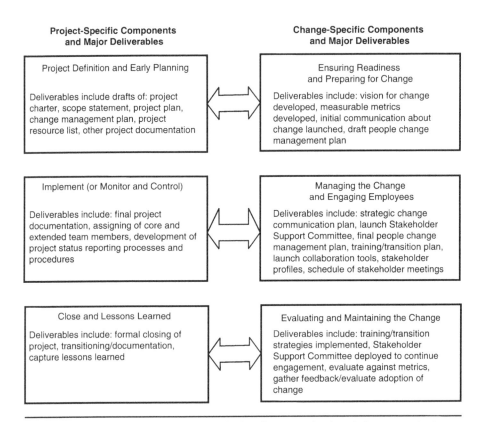

Project-Specific Components and Major Deliverables

Project Definition and Early Planning

Deliverables include drafts of: project charter, scope statement, project plan, change management plan, project resource list, other project documentation

Implement (or Monitor and Control)

Deliverables include: final project documentation, assigning of core and extended team members, development of project status reporting processes and procedures

Close and Lessons Learned

Deliverables include: formal closing of project, transitioning/documentation, capture lessons learned

Change-Specific Components and Major Deliverables

Ensuring Readiness and Preparing for Change

Deliverables include: vision for change developed, measurable metrics developed, initial communication about change launched, draft people change management plan

Managing the Change and Engaging Employees

Deliverables include: strategic change communication plan, launch Stakeholder Support Committee, final people change management plan, training/transition plan, launch collaboration tools, stakeholder profiles, schedule of stakeholder meetings

Evaluating and Maintaining the Change

Deliverables include: training/transition strategies implemented, Stakeholder Support Committee deployed to continue engagement, evaluate against metrics, gather feedback/evaluate adoption of change

Figure 7.4   More details: a framework for leading organizational change projects

For every change-focused project that my clients launch, I advise them that in addition to a project manager with experience on change initiatives, I highly recommend adding a change manager who will focus on the people-side of the initiative. This enables the project manager to focus on the technical aspects of the project overall. One client who decided to try adding a change manager to a change project determined the following:

- *Stakeholders were more engaged in the initiative and a number commented that they "felt like their opinions mattered and concerns were important."*
- *The change initiative launched with 68% of stakeholders immediately implementing the change, up from 32% just a year ago on a similar change initiative.*
- *The project manager noted that she was able to better manage the technical aspects of the project—and keep it moving*

forward—because she could rely on the change manager to deal with the people and their questions, concerns, and comments about the project.

Figure 7.5 is the organizational chart that includes the major responsibilities of the project manager and the change manager for this particular initiative.

The project manager and change manager met at least once a week—often over an early breakfast or coffee—in order to share information with each other. This enabled them to stay in alignment on the project. Many of the concerns that the change manager heard from representatives of the Stakeholder Support Committee were able to be addressed early on by the project manager and her team. The change manager enabled the project manager to stay focused

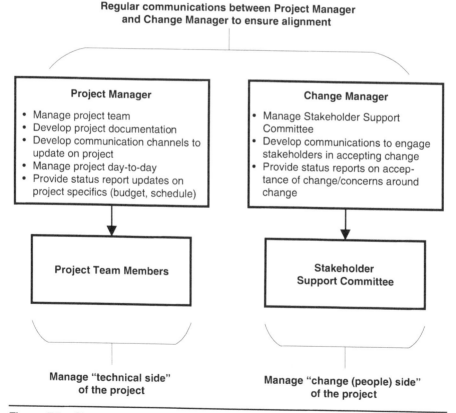

Figure 7.5   Change project organizational chart example

on moving the project forward—ensuring adherence to the schedule and staying on budget—by, effectively, running interference (along with the Stakeholder Support Committee) for those stakeholders who would be impacted by the project.

While, effectively, every project *is* a change project since every project requires something to change, organizations may choose projects that are specifically focused on change as the projects that might be led by both a project manager and a change manager. In the example in Figure 7.5, both the project manager and change manager reported up to the project sponsor. In other circumstances, the project manager reports up to the project sponsor and the change manager (assigned to complex, longer term change-focused projects) reports up to a change sponsor in the organization.

PREPARING THE ORGANIZATION FOR CHANGE

Previous chapters have focused on various aspects around preparing the organization for a change. Having only one method to prepare the organization for change is rarely sufficient for a number of reasons:

- Diversity among stakeholders requires preparing the organization in a variety of ways
- What works well in one organization will not necessarily work well in another
- The larger the organization and the larger the stakeholder pool impacted, the more ways are needed to prepare for adopting the change
- As the organization changes over time, the way the organization prepares for change will also have to change and adapt

From a strategic project management perspective, it is essential to ensure understanding of the *why, what, where, when,* and *how* of the change. This is not different from how an organization manages any other initiative that is launched. The information gathered during discussions around change, as well as the *why, what, where, when,* and *how* of the change should be delineated in a project charter. Figure 7.6 provides a template for a project charter of a change initiative.

The project charter does not, of course, predict how well the change project will be managed and implemented, but rather serves to provide confidence for senior leadership, and other stakeholders throughout the organization, that the initiative is well-thought-out. The charter will also serve as a basis for understanding potential issues or risks that may arise that should be addressed to ensure the ultimate success of the change

Change Initiative Project Charter
Proposed Change Initiative: *<name>*
Senior Leader Sponsoring the Change: *<name>*
What is the challenge or opportunity to be addressed by this change project: *<delineate opportunity or challenge here, include alignment to strategic plan, its significance to the organization>*
What is the vision for the change?

Which members of the senior leadership team support the change and why? •	**Which members of the senior leadership team do *not* support the change and why not?** •

What key resources are required to implement the change?
What communication channels are expected to be a part of the change initiative?
What indication is there that stakeholders (employees) support this change and will adopt it?
What metrics are planned to evaluate the success of the change initiative once implemented?
What are the expected risks or concerns with this change initiative?
What support or references exist that confirm this initiative is in the best interests of the organization and its people?
What is the expected, or desired, timeline for roll out of the change initiative?
Who is affected by the change initiative? (Divisions? Departments? Workgroups?)

Figure 7.6   Project charter template for complex change initiatives

initiative. If the project charter is acceptable, the next steps in preparing the organization for change are outlined in Figure 7.7.

This flow chart provides four steps to prepare the organization for change. If the project charter (Step 1) is approved, Step 2 is to review the checklist. This checklist (shown in Figure 7.2) would be customized for each organization depending upon risk factors and other concerns around change initiatives. If the checklist still indicates that the change initiative is worth moving forward, Step 3 is to share the vision for the change via

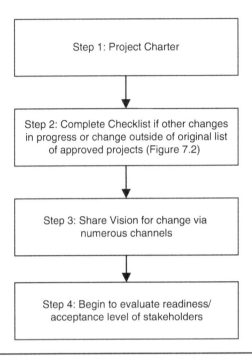

Figure 7.7   A flow chart to prepare the organization for change

numerous communication channels. The sharing of the vision is a primary part of the initial communication introducing the change to the employees *throughout the organization*. Sharing the vision also helps to make sure that everyone understands the *why* of the change.

From here, Step 4 focuses on beginning to evaluate readiness and acceptance of the change by the employees. This could be accomplished through facilitating focus group meetings, one-on-one conversations, or even via an online survey. The more time that the organizational leadership can spend to engage employees in change *before* officially starting the change project, the more likely that the change project will be successfully implemented. These conversations and data gathering to determine readiness enables structuring a much more detailed people-focused change management plan early on in the project initiative; thereby reducing risks and conflicts that will occur. However, if the organization finds that employees are just not ready for change, or will resist change, it is prudent to go back to communicating about the value and benefit (Step 3) of the change to begin to increase the comfort level, and therefore support, of the proposed change. This may entail reexamining the vision of the change.

The Chief Project Officer of a financial institution was ready to launch a change initiative that was going to centralize Project Management Office (PMO) functions. This change meant that five individual PMOs would be reduced to one central PMO. It also meant that individual employees might either have to be laid off (limited numbers) or redeployed to different areas. In particular, the Chief Project Officer knew that management would be affected. For example, only one Director of a PMO would be needed, not five. Certainly the five directors knew that their jobs were in jeopardy within the organization because they had been part of the planning process of consolidating the PMOs into one centralized PMO. The Directors as well as the Chief Project Officer reviewed the impact of creating a centralized PMO given a number of other strategic initiatives that were going on in the organization. In particular, three of the five PMOs were involved in long-term strategic initiatives. Those initiatives must keep moving forward. There was also concern that once the announcement of a centralized PMO was made to the employees of the five PMOs, there would be turnover and distractions that would impact initiatives in progress. All of this was taken into consideration, and a plan needed to be developed before communicating about the upcoming change. This included considering a number of questions that were likely to be on the minds of employees—not the least of which was who would lose their job and who would remain—and being prepared to answer those questions in a meeting with staff (or to explain why those questions could not yet be answered).

Sharing the Strategy and Vision for the Organizational Change Project

Let's explore the vision more fully. An essential component of strategy around change is a focus on sharing the vision for the change. But don't confuse the *vision for change* with the *vision for the organization;* they are two separate things. The vision for a change provides a picture of what the organization will look like *after* the change is successfully implemented. Consider it storytelling. For example:

A software company is launching a change initiative with the vision of being number one in responsiveness to the customer. Partly, this initiative is being launched due to customer complaints and perceptions that the competition is better, but also because customer service personnel have been asking to be able to be more flexible with the customer to better meet customer needs. Currently, within the orga-nization, customer service personnel are evaluated, compensated, and provided bonuses based upon the number of customer calls they handle within a given hour. This has created an environment where customers are rushed through calls to enable the customer service representative to take the next call. Additionally, current processes around taking a customer service call is very detailed and stringent. There is no room for the flexibility needed to be more responsive to the customer. In order to launch the initiative of being more responsive to customer needs, the organization will first have to undertake:

- *Restructuring the evaluation, compensation, and bonus struc-ture of customer service personnel*
- *Refining processes to enable flexibility and increased respon-siveness during customer service calls*

A senior leadership team member launches the change with the statement that, "While it won't happen overnight, we hear both you and the customer when you say that customer service must im-prove. With your support and help, we will be known as the most responsive and best customer service support organization in our industry—Number One!"

While Chapter 3 discussed how to set a vision for change and commu-nicate about that vision; in this chapter the focus will be on sharing the strategy behind a change initiative to be launched with a focus on the vision for the project. The vision for the change must be closely tied to the strategy to implement that change. Figure 7.8 provides a template utilized to prepare for the first communication around a change project. Use this template to ensure that the reason behind the change project is compel-ling and makes sense to stakeholders.

Figure 7.8 asks several questions; the responses help stakeholders un-derstand the *why* behind launching the change project. In particular, the question, *"What does the organization look like when the change is success-fully completed?"* is where the vision for the change is often captured. Tell a story here—make it exciting and full of opportunity—paint a picture of the future. The goal is to get stakeholders excited about the possibilities for the future of the organization and their futures within the organiza-tion. Consider these examples of visions for change initiatives launched at a variety of organizations:

- Engage customers who are partners with our organization, not sim-ply consumers of our services

Preparing for Initial Change Project Communications
Prepare for the initial communication about a complex, transformational change project by answering the following questions.
1.   What *external* forces are driving change? (for example: competition, customer demands, technology, regulations)
2.   What *internal* forces are driving change? (for example: leadership changes, merger/acquisition, organizational growth)
3.   How does this change benefit the organization?
4.   What is the impact to the organization if the change does not occur?
5.   How does this change benefit the individual?
6.   What is the impact to individual employees if the change does not occur?
7.   What does the organization look like when the change is successfully completed? (Describe the future.)
8.   How might individual roles and responsibilities look when the change is successfully completed?
9.   What needs to be done by the organization to support the vision for the change?
10. What needs to be done by individuals within the organization to support the vision for the change?

Figure 7.8   Template to prepare for a change project communication

- Global leadership opportunities for all employees, at all levels, to enable for continuous learning and development of individual potential
- Quicker solutions to problems through better, more effective, and confident decision making
- The long-term survival of the organization through increased innovation in products

Each of these examples provides stakeholders with a picture of what the organization wants to accomplish. They answer the question of *"what will we look like when the change is successful."*

Refer back to Step 3 of Figure 7.7: *Share vision for change via numerous channels.* Sharing the vision helps with understanding the *why* behind the change initiative. Understanding the *why* of the change permits easier adoption of the change. The *why*, however, must be explained from two perspectives:

1. The perspective of the organization (the vision)
2. The perspective of the employees (the benefit to them)

Figure 7.8 provides questions that require the leader to consider the change from the perspective of both the organization and the employees. The responses to these questions will generate better and more effective initial communication about the change.

It is essential to share this information about the change initiative via several communication channels or methods, in order to reach a broader group of stakeholders. For example, one of my clients reaches out to their employees about any major change initiative via five channels:

1. Internal website
2. Television screens throughout the organization
3. Friday Facts e-mail
4. Log in screen message on every employee's computer
5. Global voicemail from senior leadership

Through the use of each of these five channels for communicating about an upcoming change, the organization has engaged 95% of employees early on in learning about the change (up from just 50% when communications were only via e-mail).

As shared in Chapter 3, the use of storytelling is a very effective method for talking about the vision for a change in a way that does a better job of engaging employees than simply providing them with data, spreadsheets, and facts about the impact to the organization if change does not occur. While this information is certainly essential, it is *not* what engages people in and gets them excited about change! Let's go back to the example of the Chief Project Officer and the financial institution shared earlier.

> *As part of the initial communication to each PMO and its staff, the Chief Project Officer wanted to be sure that the following information, as shown in Figure 7.9, was covered.*

The Vision for the Consolidation of PMOs: *To better support a growing organization, provide a 'one-stop shop' for project management expertise and capability, and to ensure alignment of projects with strategy as well as individual divisional goals, the organization's centralized PMO will be THE driving strategic force within the organization to ensure success in all initiatives – from the simplest to the most complex.*

1. What *external* forces are driving change?

Competition; consumer as well as shareholder demand for products and services out to market at a faster pace.

2. What *internal* forces are driving change?

Organizational growth as well as an increasing demand for strategy behind projects launched to ensure alignment across the organization as well as its divisions.

3. How does this change benefit the organization?

The organization will benefit through better coordination of projects undertaken within the organization as well as through better alignment of projects with goals and objectives.

4. What is the impact to the organization if the change does not occur?

The organization will continue to duplicate project efforts across PMOs and run inefficiently, thereby not meeting time-to-market demands of consumers as well as shareholders.

5. How does this change benefit the individual?

Individual employees (project managers and other staff of the PMO) will benefit in a number of ways, including: improved utilization of resources, improved project planning by leadership, fewer projects launched, projects aligned to goals and objectives, the ability to work across the organization.

6. What is the impact to individual employees if the change does not occur?

If will be more difficult to manage the number of initiatives launched within the organization as it continues to grow. Additionally, the stress and frustration of projects 'stopping and restarting' will only continue to impact individual employees in a negative manner.

7. What does the organization look like when the change is successfully completed? (Describe the future.)

Projects launched will be aligned to strategic goals and objectives; there will be better management of projects across the organization – reducing duplication of initiatives; PMO staff will be better utilized and there will be reduced frustration and stress around projects.

8. How might individual roles and responsibilities look when the change is successfully completed?

While this has not been fully developed, it is expected that projects will be better aligned to individual's specific skills and interest areas. Additionally, there will be an increased focus on strategy and individual roles and responsibilities will enable for participation in strategic discussions. One goal of this initiative is to enable the PMO staff to define roles and responsibilities in collaboration with the Director of the PMO and the Chief Project Officer.

9. What needs to be done by the organization to support the vision for the change?

The organization needs to launch a number of smaller initiatives in order to ensure that the vision for a centralized PMO is realized. Additionally, the organization must engage those who are most impacted – the current PMOs and their staff – in structuring the overall program.

10. What needs to be done by individuals within the organization to support the vision for the change?

The staff of the current PMOs need to share their knowledge, expertise, as well as their concerns about the initiative. Through their support of the initiative, the project management function within the organization will be seen as a strategic function.

Figure 7.9   Preparing for a communication about change

To that end, he drafted the initial communication shown in Figure 7.10 that was sent via e-mail and the internal project portal as well as provided to each PMO director to share in a meeting with their staff.

As can be seen from Figure 7.10, the purpose of the initial communication was to provide some basic information prior to virtual meetings that would be held within the week. Five meetings would be held at various times—from very early morning through to later

---

To: All PMO Staff   **From:** Chief Project Officer

Re: A centralized PMO: *THE* Driving Strategic Force of the Organization for Success in all project initiatives.

I am pleased to report that the organization will be moving to a centralized PMO that will enable for:

- Improved support of a growing organization
- A 'one-stop shop' for project management experience and capability
- Alignment of projects with strategy as well as divisional goals
- Increased opportunities to work on a variety of projects across the organization for PMO staff, enabling for professional and personal growth
- Reduced projects launched, as well as eliminating the issue of projects 'stopping and restarting' due to poor planning and alignment
- PMO staff to focus more on strategic initiatives

As the organization continues to grow, managing the ever-expanding list of projects to be undertaken has become more difficult. We have heard and seen the frustrations and stress this has created for PMO staff!

We need your help however for this to be a success! We know there are a number of projects that must be completed to prepare the organization to move to a centralized PMO – this includes defining roles and responsibilities, developing a mission and vision for the PMO as well as defining best practices, processes and procedures to ensure success of the PMO. Additionally, we know you probably have more questions that will help us to further define this initiative and ensure it is planned well so that it is a success.

Please join me and the PMO Directors on any of the following dates/times via our virtual platform to learn more and to share your perspective. Of course, feel free to contact me directly with any questions or concerns that you would like to discuss prior to the upcoming meetings.

- Monday, mm/dd/yyyy at Noon EST
- Tuesday, mm/dd/yyyy at 7:00 AM EST
- Wednesday, mm/dd/yyyy at 10:00 AM EST
- Thursday, mm/dd/yyyy at 6:00 PM EST
- Friday, mm/dd/yyyy at 2:00 PM EST

Figure 7.10   Initial communication to PMO staff

in the evening—during a week's time period to accommodate various time zones as well as schedules. This would ensure that everyone had a chance to hear the message from the Chief Project Officer and had the opportunity to ask any questions about the initiative.

The Chief Project Officer started each of the virtual meetings with the following statements:

> *"Imagine if we were **the** driving strategic force behind every project that gets accomplished in the organization. Imagine if we **always** launched the right projects at the right time and they were aligned to strategic goals and objectives. Imagine working in an environment where you were not stressed or frustrated. And, imagine not only using your current skills, knowledge, and expertise but having the opportunity to continue to develop professionally and personally. That is where we are heading and we need your help to get there!"*

Every change launched within an organization must have some strategy around how to launch and implement that change. The most effective organizations—those who *do change well*—have a strategy behind launching change. That strategy includes a project management approach as well as standards around communicating about the change and managing impacts to people, processes, and technology. Table 7.1 provides some key areas of strategic consideration to aid in ensuring that change initiatives are well-thought-out before launching, well-implemented when launched, and more likely to be successful—meaning the change *sticks*—overall when rolled out to the organization.

> *After gathering feedback from employees over a three year time period, a project management consulting firm realized that change initiatives were launched in a haphazard manner. There was no real thought behind leadership's support of the change and whether staff had the competencies to make the change work over the long term, or how processes, procedures, or even technology had to be adapted for the change. Through the use of a checklist with information similar to what is in Table 7.1, the organization did a better job in launching the right change initiatives at the right time and for the right reasons. This accomplished a number of benefits for the organization and its employees, including:*
>
> *   *Less stress among employees when change initiatives were launched*

Table 7.1 Key strategic areas to increase the success of change

Strategic Area	What to Consider...
Leadership	• Does leadership support and embrace change in the organization? • Does leadership approach change from the perspective of the good for the organization *as well as* the value to the employees?
Management (one level down from leadership)	• Does middle management embrace change? • Does middle management look for areas of continuous improvement opportunity without prompting from leadership?
Communications	• Are there numerous standards for communicating within the organization? • Is there regular top down, bottom up, as well as horizontal communications? • Are there protocols for communications that ensure consistency, efficiency, and sharing of information? • Is there a format, or formula, for the initial communication about the change initiative?
Change project team	• Are individuals selected to serve on the change project team skilled and dedicated to the role? • Are individuals selected to serve considered to be champions of change? • Do individuals selected represent diversity in the organization? • Do individuals selected to serve have informal power and the ability to influence?
Competencies	• Are competencies aligned to strategic goals? • Is there an understanding of key competencies and where gaps exist? • Does the organization track competencies and align to needs?
Transition	• Are transition plans a required part of every change initiative? • Are transition needs taken into account as part of the overall planning process—including resource needs, budget, etc.
Technology	• Does the organization consider how technology may impact or benefit the change initiative? • Does the organization understand, or have a process for evaluating, any technology needs of the change initiative? • Are technology plans a part of change initiatives?
Processes and Procedures	• Are processes and procedures regularly updated within the organization (continuous improvement)? • Are reward systems in place to acknowledge changes to how the work gets done as well as to enforce adoption of the change? • Are processes and procedures considered in early planning of change initiatives?
Training/ Development Plans	• Are training plans a component of every change initiative? • Does the organization evaluate the current skills and knowledge of employees to understand gaps in knowledge that may impact a change? • Does the organization consider future skills needs and build that into strategic plans?

- *Increased success of change initiatives (since the right initiatives were launched at the right time)*
- *Improved alignment of change projects with resources available as well as time available*
- *Increased acceptance of change due to forethought in how to roll out and training needed to increase comfort levels of employees*

DEVELOPING YOUR PEOPLE CHANGE MANAGEMENT PLAN

Table 7.2 distinguishes the plan to manage changes to a project from a people change management plan. Often, project managers will refer to their people change management plan when referring to a plan to manage changes to the project.

The plan that shares the process for managing changes requested to the project should include information that ties back to the people change management plan, should that requested change further impact a particular stakeholder group. For example, let's assume a senior stakeholder requests that a different technology than initially proposed is implemented as part of a larger change initiative. This is

Table 7.2 Difference between a plan to manage changes to a project versus a people change management plan

Plan to Manage Changes to Project	This particular plan enables the project manager to have a process to manage changes requested to the project. These are changes that impact scope, timeline, resources, etc. This plan captures information such as: • What the change is • Why it needs to happen • The impact of the change • The cost of implementing the change • Whether or not the change is approved This plan is managed by the Project Manager.
Change Management Plan	This particular plan provides a process and foundation for managing people's expectations of change. It enables individuals to move through the stages of adapting to change (Figure 2.3) and provides a solid foundation for how stakeholders at large will be engaged in the change. This plan is managed by the Change Manager.

a newly released technology and something that will add significant value within the organization. This change has been approved. This technology will replace the initial technology selected. The use of this technology requires training to be redeveloped and will impact another group of stakeholders due to the complexity of the technology. The change request detailing the requested and approved change is shown in Figure 7.11.

Change Request Form	
Project Name: Process and Technology Upgrade Initiative	*Date of Request*: xx xx, xxxx
Project Manager: JSA	*Individual Requesting Change*: GHK
Description and Reason for Change: New technology has just been launched by XYZ Software Partners. This technology has been launched earlier than expected and preliminary testing shows that it is compatible to our systems and will enable for more effective implementation of the project. It is expected that this technology will better meet our needs than the initial technology proposed (which was proposed because this technology was unavailable).	
Impact of Change if Approved:	• Budget impact: + $150,000 • Schedule impact: +3 weeks • Resource requirements: 2 additional contractors at $65,000/each • People (stakeholder) impact: Stakeholder group B, initially not impacted by the project, is now going to need to be involved in the new technology and, in particular, integrating the technology with their other systems. • Other impact: Training is already in development for initial technology selected; will require restarting of training component of project. + $45,500 • Stakeholder Support Committee: Requires understanding of new technology and its benefit to share within the organization.
Decision Made and by Whom:	APPROVED on xx xx, xxxx by Project Sponsor
Reason for Approval:	Technology better supports initial goal of project. Project would have been relaunched – Phase 2 – once this software became available.

Figure 7.11   Change request example

In the example change request form shown in Figure 7.11, there are two specifics areas of impact that are people-focused (stakeholder) that would need to be addressed by the change manager and his team. First is the fact that Stakeholder Group B will now need to be involved in the initiative. This means that this group, while they are aware of the project overall, will now be more closely involved in the project. This requires getting this group engaged in the initiative through understanding how it impacts them both as a group and individually. Second, the Stakeholder Support Committee will need to understand more about the new technology and its benefits within the organization. They will need to share information about this change and its impact with the greater stakeholder pool. It can be assumed that if training is already in development for the initial chosen technology, that work will, effectively, be thrown to the side. New training will need to be developed. If any stakeholders were involved in shaping and testing any training, they will need to be communicated with and re-engaged so they will continue to support development of training. Once the change request is approved, the project manager would work closely with the change manager to ensure that they are collaborating around incorporating this change into the project. Certainly the project manager would have involved the change manager in evaluating the impact of the proposed change prior to submitting to the sponsor for approval, rejection, or deferment.

The purpose of the people change management plan is also to move stakeholders through the stages of adopting change. Recall Figure 2.3: Stages of adapting to change for individuals. The people change management plan, in conjunction with the communication plan discussed later in this chapter, should enable for moving employees from Stage 1 through to Stage 4.

Figures 7.12a and 7.12b provides a template for a change management plan. This template is available as a downloadable file from the Web Added Value™ Resource Center at www.jrosspub.com/wav.

The template shown in Figures 7.12a and 7.12b is focused at a high level on engaging the people—stakeholders, internal or external—in the change. For every change initiative launched in the organization—even the simplest—a people change management plan should be developed. Much of this information may be transferable for future change initiative use. For example, information around the purpose of the communications as well as the types of communications utilized (Figure 7.12b) is likely to be fairly consistent within the organization.

CHANGE MANAGEMENT PLAN	
Project: *<name>*	Date: *<of plan>*
Overview:	*<Include here the "why" of the project; mission and vision for the change; impact of not implementing the change.>*
Team Members:	*<Include: sponsor, project manager, change manager, project team members, Stakeholder Support Committee members, etc.>*
Current and Future State	
Current State:	*<Describe, at a high level, the current state. What processes need to change and who needs to be involved in making the changes?>*
Future State:	*<Describe the ideal future state. What will processes look like at a high level, what skills and knowledge will be needed by stakeholders, what documentation is needed, what else will change in the organization?>*
Potential Impacts:	*<List potential impacts: stakeholders, customers, org structure, divisions/departments, budget impact, regulations/laws, etc.>*

Stakeholders Impacted (by group or department, list individually if necessary)			
Name:	*Awareness of Need to Change:*	*Support Needed for Change to be Successful:*	*Ability to Influence Initiative:*
Name:	*Awareness of Need to Change:*	*Support Needed for Change to be Successful:*	*Ability to Influence Initiative:*
Name:	*Awareness of Need to Change:*	*Support Needed for Change to be Successful:*	*Ability to Influence Initiative:*

Figure 7.12a   People change management plan template

CHANGE MANAGEMENT PLAN (continued)		
Communications		
Communication Responsibilities:	Change Manager *<Include specific responsibilities of change manager for communications.>*	Stakeholder Support Committee *<Include specific responsibilities of Stakeholder Support Comm. for communications.>*
Purpose of Communications with Stakeholders:	*<What is the purpose of stakeholder communications? For example, engagement, buy-in, support, etc.>*	
Types of Communications to be Utilized	*<Include information about acceptable types/modes of communication – such as email, posters, focus groups, Stakeholder Support Committee, project website, etc. This section will link back to the full communication plan – see Figure 2.2 for an example.>*	
Training and Documentation		
<Delineate here training and documentation needs for ensuring that the change initiative is adopted and "sticks" over the long term. Include suggested job aids, delivery method for training as well as specific stakeholder training needs. This should be an overview only as this section would link to a detailed formal training plan.>		
Post-Change Implementation Needs		
Dept/Team	Resources	Time Period
<What dept. or team is needed for post-change support and what is their role, e.g., Tech support, training.>	*<Who specifically – names of individuals – are needed?>*	*<What is the time period needed for support? From when to when?>*

Appendices *<include in Appendices any necessary documents such as training plans, project plan, Stakeholder Support Committee information, change management team org structure, etc.>*

Figure 7.12b   People change management plan template (continued)

For one particular organization, every people change management plan developed includes the following types, or modes, of communications to be utilized to engage stakeholders in change:

- *Posters*
- *Internal website home page*
- *Monday morning manager's message*
- *Friday newsletter*

The purpose of the people change management plan is to outline at a high level who is involved and what methods will be used to engage stakeholders in the change. When developing the change management plan, consider the goal of ensuring that stakeholders will understand:

- The vision for the change initiative
- The business goals of the change initiative
- The value of the change initiative to stakeholders (beyond business value)
- Activities or tasks associated with implementing the change initiative
- Training and other documentation or job aids that will be provided to ensure adoption
- The ongoing progress and check-in points (consider check-in points as the opportunity to get feedback on the change from the stakeholders)
- Points of contact for reaching out to the individuals responsible (e.g., Stakeholder Support Committee, change leader, change manager) for leading people through change

As a best practice, the people change management plan should be developed in collaboration with the Stakeholder Support Committee members (or a subset) and with the change team members. While the change manager is responsible for the plan overall, other input is needed to ensure a robust plan that meets the needs of the stakeholders.

As a best practice, any tasks associated with the change management plan should be incorporated into the overall project plan. This enables a more complex overview of the project and ensures collaboration between the project team and the change team.

The people change management plan provides an overview of how the organization will move from the current state (what is) to the future state (what will be—the vision). Refer to Figure 1.4: Stages of change—today to tomorrow. An understanding of the current state—what the organization looks like, acts like, and how it works today—is necessary to understand the impact of the change project on the stakeholders. The current state is the basis for determining the robustness necessary for the people change management plan. The transition stage identifies what needs to change, i.e., skills of employees, processes with which the work gets done, office locations, etc., in order to accomplish the future state—the organization of tomorrow. As has been stated earlier in this book, the more complex the change (the more changes required of the people within the organization) the sooner the people change management plan must be activated. The gap between the current and future state determines how early, *prior to project start*, to begin communicating about the initiative as well as the number and types of communications that will be required.

> *My work with a west coast-based national client required the people engagement part of a complex change initiative to begin two and a half months prior to the actual work of the project beginning. The launched project required the opening of several new locations and closing, and relocating employees, from two other locations. Before the work of the project began, it was important to ensure that employees—especially those who would need to relocate—understood why the change was happening and the value of the change to them. We also knew that there would be questions around: what happens if an employee chooses not to move, what moving expenses would be covered, what happens if an employee already owns a home, etc. It was important to have an understanding of the details before engaging the employees in the change.*

Communication Planning

Chapter 3 shared some best practices for communicating around change and developing a communication strategy. This section will further discuss communications with a focus on aligning communications about the project overall with communications around engaging stakeholders in the change. There are two trains of thoughts around communications for change-focused projects:

1. Have *one* communication plan that includes information about communicating on the project as well as communications to engage stakeholders in change.
2. Have *two* communications plans—one focused on project-only communications and the other focused on engaging stakeholders in change so that they more easily and readily adopt it.

I prefer *two* communication plans. Based on past experiences, two communication plans tend to enable the right people to stay focused on the right areas. For example:

> *One of my clients has two communication teams for each change project that is launched. One team, which includes the Stakeholder Support Committee, is focused on communicating and sharing information with stakeholders. The other team is focused on communicating up to leadership about the project status overall as well as communicating among and between contractors and team members about the project. As part of this team's communications to leadership, they will coordinate with the lead of the change communications to understand what percentage of individuals are engaged in the project and challenges that may have arisen around engaging individuals.*

A strategic change communication plan is focused on engaging stakeholders from where they currently exist in terms of being engaged in change. If an organization has failed at numerous previous change initiatives, communication planning should start early on—ideally before the launch of the initiative. An initial communication that would be part of the plan would focus on acknowledging less than successful past change initiatives and sharing why this change initiative will be different (effectively, what were the lessons learned last time). Organizations that have undergone successful past change initiatives may focus on reminding individuals of those successes, how they have contributed to those past successes, and how they can contribute to future success.

Recall the example of a partially completed communication plan in Chapter 2 (Figure 2.2). The column titled, *Overall Timing* (project phase), should be aligned to the project management phases, or steps, used within the organization. This information ensures that communications for the project and communications for engaging in change, while separate, are not overwhelming for any particular group at any point in time. For example:

> *JKL Corporation is launching a 16-month initiative that will impact all processes within three divisions. The project manager developed a communication plan that was focused on updating the leadership team on the initiative as well as ensuring that stakeholders were aware of times and dates they would need to be available to perform a number of tasks, such as providing input on current*

process issues, validating redesigned processes, and participating in testing initiatives. The change manager developed a communication plan that initially focused on explaining why processes were changing, sharing the vision for the change, as well as noting individuals within each of the divisions who had already improved upon processes. This communication also provided information about the Stakeholder Support Committee and how to reach out to them with questions, concerns, or ideas. A second e-mail to be sent would ask for volunteers to assist in documenting processes, validating processes, and participating in testing. This e-mail from the change team would be sent and a follow-up meeting held prior to the project manager's team sending out information regarding dates/times for stakeholder help.

This simple example shows coordination between the project manager and the change manager over communications. Coordination of communication ensures that stakeholders are not overwhelmed by the number of communications they receive and that they receive the right communications at the right time, to ensure overall change project success. In this case, the first communication to be sent is to ensure understanding of why the project is happening and to engage employees in talking about the project. This would be done prior to asking for help in participating on the project. It is essential to engage individuals early in the change *prior* to asking for their support on the initiative.

As part of communication planning, change managers should gather some information that will enable them to more effectively address questions that will likely arise about the change. Table 7.3 provides several questions to ask during initial information or data gathering to ensure that the change manager as well as the Stakeholder Support Committee can answer the many questions that will likely arise as the change project is introduced to employees.

The better prepared that the change manager, Stakeholder Support Committee, and others who need to communicate are in predicting questions and providing responses, the more likely that employees can be made more comfortable about the change about to occur.

A law firm was expanding and branching out to other areas of law through a number of mergers and acquisitions. One of the partners, who had worked with change managers at another law firm, brought in a change manager to help frame the initial communications about the merger and acquisition. Responses to the questions

Table 7.3   Questions to consider prior to conversations around change

Information Gathering—Answering Common Questions	
Question...	*Consider in your response...*
What problems are being addressed by this change initiative?	• How prevalent are the problems? • Can the problems be traced back to impacting the productivity of employees? • Will solving the problems enable for achieving personal goals (not just organizational goals)?
What do we expect to be different when the change is implemented?	• What, specifically, has to change? • What is the vision for the change?
How does this change link, or align, to other changes happening now in the organization?	• Why does this change have to happen now? • Is this change because of another change already in progress, or does this change support another change that has to happen?
How will this change help us to achieve our goals and objectives in our division or department?	• Can this change be aligned to specific division or department goals? • Will it help the division or department solve a problem that is impacting achieving a goal? • Will it help the division or department to achieve a strategic goal?
What impact will this change have on: how I work, overall processes and procedures, employee roles and responsibilities, compensation, and organization structure?	• While specifics are not needed early on, consider—at a high level—potential impact in any number of areas that may be of concern. Is the impact mild? Moderate? Significant? • What support is needed to help determine the impact in certain areas? How can employees help?
What happens if the organization does not make this change?	• Consider not only organizational impacts, but individual impacts also (e.g., it will be more difficult for employees to get work done due to cumbersome processes).

were prepared (as they were expected from attorneys in the firm). These questions may include:

- How will the merger and acquisition impact potential promotions to partner roles?
- How will seniority be determined?
- How will clients be shared, if at all?

Responses were prepared to questions expected from support staff. These questions may include:

- How would support staff be assigned to support attorneys?
- Would any support staff lose their jobs due to the merger?

Figure 7.13 provides a brief breakdown of one question to dive deeper into understanding what stakeholders may want to know regarding a change initiative.

Let's assume one of the potential questions for which a response is being prepared is: "What problem is being addressed by this change initiative?" First, the change manager might consider whether the problem is prevalent in the organization. If it is prevalent, this is an easier sell to stakeholders. The change manager may be able to mention that the problem was brought to their attention by end users (the stakeholders themselves) and is being addressed by leadership, or they may link the problem to a difficulty that stakeholders may be having in performing their role and therefore, once addressed, will make the job easier. Alternatively, let's assume that the problem is not prevalent. In this situation, the change manager will need to spend more time selling the fact that there is a problem *before* he can talk about why the problem needs to be solved. Here, as shown in Figure 7.13, the change manager may focus on demonstrating the impact of the problem on the organization as well as on the individual.

Not only does this type of information gathering and preparation of responses to potential questions help in framing communications with stakeholders, but it is a great way to have conversations around change.

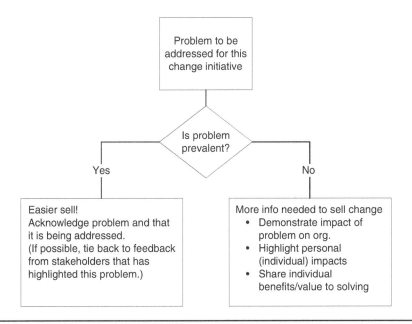

Figure 7.13   Example of preparing to address questions from stakeholders

For one of my clients, the Stakeholder Support Committee assigned to represent the greater employee group held mini breakfast sessions with coffee and pastries twice a week for a three-week period before the project officially kicked off. Each session answered a question. While some questions were prepared ahead of time; other questions were gathered by the employees through an anonymous survey. The survey asked one question: "Given what you have heard to date about the change initiative from the CEO, what questions may we answer for you?" Each of the breakfast sessions were recorded and posted on the internal site so that those who could not attend the sessions could still learn about the project.

In addition to gathering information about the types of questions that may arise, it is of value to create an overall plan about each stakeholder group that will help shape communications and conversations with them. Recall Figure 5.2: Stakeholder analysis template. Figure 7.14 provides another example of a document that may be used to better understand stakeholders. This template is also available as a downloadable file from the Web Added Value™ Resource Center at www.jrosspub.com/wav.

Figure 7.14 focuses on looking at stakeholders from the perspective of an entire group or department level. However, since it is unlikely that an entire department is working on components of a change project, you may want to highlight individuals who will have to contribute to or support the project in some way. Certainly, the template shown in Figure 7.14 may completely replace the template in Figure 5.2, or vice versa. The reader should determine which template is most useful for analyzing and managing stakeholder perceptions and communicating about the change initiative.

Use the templates in Figure 7.14 and Figure 5.2 as follows: Figure 7.14 is the primary template—it focuses on stakeholders in groups (such as work teams or departments). Figure 5.2 is used as a secondary template and is focused on the individual within a particular stakeholder group. For example, let's consider Finance (first line in Figure 7.14). After this quick analysis, discuss options with the Head of Finance. Prior to doing so, prepare data to populate Figure 5.2. For example:

- *Stakeholder: Finance Director*
- *Impact: High*
- *How are they impacted: Need to assign individuals within Finance to assist project team in evaluating current processes to determine gaps based on changes being made*

Individual Stakeholder	Impact from change	Concerns that may exist Validated? (y/n)	What is needed from the individual	Best mode of communication
Finance	Change processes for how invoices are processed	Vendor contracts may impact process changes (y) End of year closing may impact ability to change processes in a timely manner (y)	Support in documenting current processes. Support in designing new/updated processes.	Sharing information at Finance dept. meetings.
Procurement	Will need to work with legal to change contracts for vendors to capture new process	Contracts were changed less than 6 months ago; therefore may be frustration among procurement staff that contracts are changing again (n)	Analysis of current contracts to understand where changes will need to occur. Plan to update current contracts in effect with vendors	Sharing information at Procurement dept. meetings.
Legal	Will need to help Procurement with changes to current contracts as well as developing new contracts.	Already involved in a number of lawsuits that will impact their ability to take the time to assist procurement; which may cause delays in this change initiative (y)	Work collaborative with Procurement in updating/changing contracts	Via email

Figure 7.14   Template to capture information about stakeholders

- *Effort required: significant as current processes are not well documented*
- *Perception of change: likely to be some frustration because the Finance Director is under pressure to get books closed for year; additionally, the time investment in documenting current processes will take those involved "off-line" from other finance work*

Ensuring People Are Ready for Change

Although communications are essential for *any* type of project launched within the organization, communications are particularly of essence in change initiatives. Most project communications are primarily focused on sharing information about the project, with key stakeholders and project sponsors communicating between and among team members and reporting on individual team members' task status. There may be a company-wide e-mail that goes out that shares information about a particular project in the works or that asks for help in documenting requirements or testing before go-live. However, if there is agreement that *every project represents change for someone somewhere* in the organization, and therefore it is necessary to engage those individuals in change to ensure success of the initiative, then communications become much more essential than the usual project communications. Additionally, similar to the necessity to evaluate project communications, it is absolutely essential to evaluate communications on change. Consider whether change-focused communications are going well by:

- Providing stakeholders with information they need and want when they need and want it
- Adapting to changing needs as the project progresses and more is learned
- Continuing to engage individuals who may still be resistant *even if* it appears that the majority champion the initiative
- Regularly including a reminder about the reason for the initiative and the vision for the change

The most effectively designed and planned communications should help to make sure that people are ready for change. If they aren't ready, that is an indication that the project should not be started until such time as the people are made ready. Chapter 4 provided best practices for helping ready the organization for change, as well as some best practices for ensuring people are ready for change. Here, we'll build on Chapter 4 with a focus on information gathering, communication strategies, and the deployment of change teams to ensure readiness for change.

For those readers who work within an organization, it is easier to tell, over time, the readiness of people for change. The people in the organization become known—relationships are established, trust is built, conversations happen regularly that indicate (to those who listen) whether or not a particular change initiative will be successful if launched, or if significant engagement of the people must occur to enable for that success. For those readers who are consultants and work with a variety of organizations as I do, the task is a bit more difficult to accomplish. Time must be spent building relationships and establishing trust and thus time must always be allocated to engaging individuals and ensuring readiness of change *prior* to launching the change initiative.

A friend of mine shared the following story: "PMOs are notoriously difficult to implement from scratch. Why? There are two main reasons. The first is that the majority of executives, still today, do not realize what benefits can come from a good, centralized project and portfolio management and governance practice. The second is that instituting a PMO, no matter how small, inevitably will change the company's culture. Influencing executives and changing culture are two of the more intractable challenges in business. Realizing this up front and engaging in a comprehensive stakeholder management strategy from the outset is the best way to begin a wide-ranging change management initiative such as a PMO. The best approach is to never believe that your stakeholder management work is done."

Gathering information about the stakeholders prior to designing communications and having initial discussions (Figure 7.14), enables better engagement of stakeholders—as a group and individually. Of all the change initiatives on which I have worked, as well as my friend who contributed to this story, PMOs stand out as difficult in many ways; in particular, when change initiatives are launched from high up in the organization—the executive level—without any regard for what is truly involved in undertaking that change.

As an aside, PMOs, when effectively used within the organization, are often the home for change management best practices.

Overall, organizations that have made change a normal experience often have people who are ready for change. This is because change is such a frequent occurrence that employees expect change to happen. People who are ready for change:

- Offer up new ideas and embrace new ideas presented to them
- Collaborate more effectively across the organization/across divisions to accomplish goals
- Share knowledge
- Are encouraged to take risks
- Are encouraged to solve problems
- Are competent at leading change

Even when it appears that people within the organization are ready for change, test that theory! Understanding more about the perceptions people have of an upcoming change *before* communicating about the change will allow you to tailor communications to ensure that they provide people with the information they need to accept the change. Consider this example:

A global resort and spa is known for providing the highest level of customer service—at a price to match. They cater to wealthy individuals, often hosting actors, politicians, and millionaires. However, over the years, revenues and profitability have been decreasing as competition has increased for the attentions of their clientele. The resort and spa leadership believes that they can attract more individuals if they reduce their services to lower the cost of staying with them. This would also increase their profitability. Initially, the belief was that the staff would be unconcerned with this change as it would mean, effectively, less catering to clientele. The e-mail that went out to all employees, as well as the handouts left in employee areas, shared that the organization would be moving forward with a large change initiative that would enable for expanding the customer base through changing and/or reducing services to reduce costs of staying at the resort.

Two directors, however, were concerned about this initiative. They took it upon themselves to have informal conversations with various staff to get their opinions about the e-mail that was sent by leadership. It was apparent that feedback about the change was less than positive.

Let's pause with this client story for a moment. Informal conversations were initiated *not* by leadership, who were ready to move forward with launching the change project—rather, they were initiated by two of the directors—the director of operations and one of the resort directors. They were initiated based on the following reasons:

- Concern with a move to provide less services to clientele
- Concern with the impact that the change may have on jobs
- A perception that the change was not well-thought-out but rather was reactionary

For these reasons, as well as because there has been no major change initiatives launched within the last seven years, the directors felt it necessary to do a bit of research *prior* to bringing their concerns to leadership. Let's return to the client story.

After a number of informal conversations, both directors approached leadership with a summary of their conversations (shown in Figure 7.15, partial summary only; 25 employees had been interviewed as part of the informal conversations).

Role	Concerns	Ideas to Solve Problem
Housekeeper	• Loss of jobs • Managing angry clients	• Offer frequent stay points for individuals who don't need/want room cleaned daily • Provide pricing based on services desired
Operations Manager	• Loss of jobs • Perception that client is not as important • Further reduction in profitability	• Survey clients to determine services desired/services no longer desired • Look at other ways to increase profitability and attract other clientele
Security Admin	• Loss of jobs	• Look at other ways to increase profitability and attract other clientele
Front Desk Reception	• Managing angry clients	• None offered
Golf Pro	• Managing angry clients	• Research what other resorts and spas have done to increase revenues
Restaurant Server	• Managing angry clients • Loss of jobs	• None offered, unsure of what is actually changing and wants to learn more first

Figure 7.15 Summary of concerns with proposed change based on informal conversations with employees

As can be seen in Figure 7.15, there were two primary concerns (these concerns were reflected in the other interview data not captured in this figure)—loss of jobs and managing angry clients. While some of the individuals interviewed offered up other ideas to meet the goal of increased revenue and profitability, the majority of respondents (not reflected in Figure 7.15) felt they just did not have enough information to give a solution.

The directors shared this information (keeping individual names off the list of those interviewed) with leadership. Bottom line—a decision was made to share more information with staff before launching any change and asking for small groups to offer up solutions that would help to increase revenue and profitability.

Had senior leadership moved forward with a change to reduce services, it is unlikely that the initiative would have been successful. People in the organization were not ready for change. They had limited understanding of why the organization was changing (there was no vision for change shared), had limited experience with managing through change, and had great fear about the impact of the change on them personally. Additionally, pushing forward with an initiative such as this client example when employees may not be ready could result in other challenges for senior leadership—such as employees expressing dislike of the change to long-term clientele with whom they may have established close working relationships.

The idea of informal and formal conversations is a great way to engage people in change prior to actually launching the change. This may be approached in any number of ways: through surveys, one-on-one meetings, conversations over lunch or in a cafeteria, via department meetings, focus groups, or in any other number of ways. Certainly in order to get the most information from these conversations, the organization must share the idea for the change—even if at a high level. Often these early conversations provide an understanding of readiness for change within the organization.

As a best practice, prior to launching any change initiative, ensure that the people of the organization—those individuals who accomplish the day-to-day work—are ready for change. Start informal and formal conversations, listen to employees' perspectives, and aspire to understand where they see challenges. Use that information to structure communications and the change initiative in a way that will engage and get the people involved.

Developing and Deploying Your Change Teams

Change teams were introduced in Chapter 5. In this chapter, more will be shared about change teams with a focus on collaboration between change teams and project teams.

> *The change team is a group of individuals from throughout the organization who are responsible for supporting the change component of the initiative and ensuring adoption of that change.*

The change team supports the change manager in managing stakeholder perceptions and adoption of the change initiative. Change agents (as discussed in Chapter 5) may certainly be an extension of change team membership, as might be the Stakeholder Support Committee. The change team may be appointed by the change manager or by the change leader. The ideal core change team size is five to eight people. Similar to project teams, too large a core team can be unmanageable. The ideal size of five to eight does not include extended change team members such as change agents (champions) or Stakeholder Support Committee members. Additionally, this number is simply an *ideal* and not a definitive. The number of change team members is going to vary from one organization to another. Table 7.4 provides a number of items to consider when appointing change team members.

Table 7.4   Considerations when appointing change team members

• Individuals who represent those areas of the business most impacted by the change initiative • A mix of supervisors as well as individual contributors • Individuals with specific expertise such as training, process improvement, etc.	• Diverse personalities, e.g., creative individuals, "can do" individuals • One or two individuals who bring an outside perspective (not impacted by the change) • Individuals with strong relationships, seen as "go to" people, have built trust

Similar to challenges or barriers faced by project teams—those individuals who are completing the tasks required to accomplish project objectives—there are a number of challenges or barriers that can be common among change teams. Table 7.5 provides a list of potential challenges or barriers that are faced by change teams as well as some ways to avoid or address the challenge.

Most of the challenges can be avoided when the following is accomplished *prior* to getting the change team to work engaging and conversing with stakeholders:

- Ensure clear roles and responsibilities
- Cooperate with the change team to develop processes and procedures for collaboration
- Evaluate stakeholders and their needs
- Develop a comprehensive communication strategy with regular check-ins to ensure effectiveness
- Ensure channels to enable communication and feedback loops between change team and project team members

Introduce the change team to the stakeholders prior to scheduling any stakeholder meetings. The introduction should include information such as:

- The names and contact information for each member of the change team
- Roles and responsibilities of change team members
- How the change team will communicate with stakeholders
- How stakeholders can communicate/reach out to change team members

If the change initiative also includes a Stakeholder Support Committee, be sure to distinguish the work of that committee from the change team's efforts. Figure 7.16 provides an organization chart (with responsibilities

Table 7.5   Challenges faced by change teams and ways to avoid/address the challenge

Challenge	How to Avoid or Address
Lack of empowerment and/or engagement	• Ensure understanding of vision for change initiative • Be clear about responsibilities of serving on the team • Ensure that the team has processes for making decisions, solving problems, and sharing the work of communicating about the change
Insufficient or too much communications	• Ensure a strategy and plan is developed around communications • Balance communications from the change team with communications from the project team (consider a role that oversees all communications) • Evaluate the effectiveness of communications at regular intervals
Limited communication channels	• Ensure that a variety of communication channels or modes are established to reach the largest group of individuals
Misunderstanding of stakeholder needs	• Ensure a stakeholder analysis is completed • Understand why the change should matter to individual stakeholders—what's in it for them? • Know what is important for stakeholders
No sharing of information or feedback	• Develop processes for sharing information about stakeholders between change team members • Ensure there is a process and method to capture and share feedback on the change between the change team and the project team
Task of engaging stakeholders seems too large to accomplish	• Break down the efforts of the change team into smaller components with clear goals for each component
No success in engaging stakeholders	• Step back and re-evaluate what stakeholders need • Re-evaluate the stakeholder analysis conducted, or, if not conducted—take some time to develop a stakeholder analysis • For larger initiatives, assign stakeholder groups to particular change team members
Lack of coordination among change team members	• Ensure clear roles and responsibilities on the change team • Ensure processes and best practices around communicating between change team members • Consider the use of an internal site to capture information to share between change team members
Project team and change team not collaborating or working together	• Ensure coordination between the two teams led by the project manager and the change manager • Schedule and facilitate regular meetings between both groups to ensure sharing of information

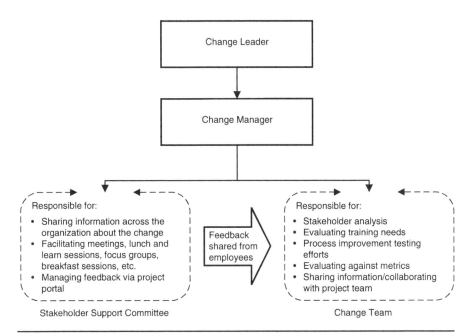

Figure 7.16   Organization chart: Stakeholder Support Committee and change team

highlighted at a high level) for both a Stakeholder Support Committee and a change team for work on a large, complex organization-wide change initiative.

On a weekly basis, a representative of the Stakeholder Support Committee met with a representative of the change team to share feedback gathered from stakeholders. The change team used this information to determine whether stakeholder needs were being met. For example, at one point early in the project, several stakeholders commented to the Stakeholder Support Committee that they were disappointed that the only training that would be provided would be face-to-face training options. Given the travel schedules of many of the stakeholders impacted by the change initiative, it was felt that training needed to be provided in a number of formats, not just face-to-face. The change team used this information and restructured the training for the change initiative. The team developed many methods for training stakeholders, including face-to-face, virtual, e-learning, and *how-to* sheets. The change team then shared this information back with the Stakeholder Support Committee to announce to the stakeholders.

The Roles and Responsibilities of Change Teams

Let's focus a little more about roles and responsibilities of change teams. Just as it is necessary to be clear about roles and responsibilities of project teams, it is essential to be clear about roles and responsibilities of change teams. Without clarity, challenges in implementing the change initiative increase and the success of the change initiative declines. Table 7.6 provides a variety of potential roles of change team members and responsibilities associated with those roles.

Table 7.6 provides only a partial list of potential roles of change team members. An Internet search will return a list of any number of roles (titles) for change team members. Consider aligning roles with common roles already in place within the organization or in alignment with project management team roles being utilized.

> *One organization that I work with on every complex change initiative enables the change team members to select their own roles based on what interests them and on their expertise and skills.*

Getting a Broader Group of Stakeholders Involved

The creation and deployment of change teams, change agents, and Stakeholder Support Committees allows for a broader group of stakeholders involved in a change initiative. This does not mean to imply that simply adding individuals to serve on a change team—as a change agent or on a Stakeholder Support Committee—is sufficient. It is necessary to ensure that roles and responsibilities are clearly defined and there are meaningful tasks for each individual serving in a change-focused role.

> *A new leader in an organization noticed that each change initiative engaged a large group of stakeholders in the change. However, he also noticed that the individuals had no clear responsibilities regarding the change. At the start of one initiative, the project manager shared communications with this group of stakeholders and asked them to then share throughout the organization about the change. Some stakeholders did so, others did not. Worse yet, the conversation between any one of these stakeholders and others they updated about the change was significantly different. The right information was not being shared! Prior to the launch of the next change initiative, the leader worked with his senior staff, human resources, and individual contributors who have worked on change initiatives to develop a list of roles and responsibilities for individuals serving on*

Table 7.6   Potential roles and responsibilities of change team members

Change Team Role	Responsibilities of that Role
Change leader	• Develop the strategy for the change to ensure success of the change • Sponsor the change initiative as a champion of change
Change manager	• Oversee the work of the change team • Collaborate with project manager • Develop a strategy for change • Lead development of strategy for communications, training, testing, transitioning, etc.
Change consultant	• Provide expertise on change • Provide assistance to change manager
Change administrator	• Provide administrative support to change manager and other change team members • Schedule/manage logistics for meetings
Communication specialist	• Provide input to communication strategy • Determine/develop/launch a variety of communication channels • Ensure use of communication channels • Manage overall communication plan
Training specialist	• Responsible for managing to training plan • Developing/sourcing options for training on change
Change analyst	• Provide analysis of stakeholders' readiness for change • Provide information on barriers or challenges to change • Assess impact of change on the organization
Organizational change management consultant	• Develop and execute organizational change management plans • Push forward with positive change within organization • Analyze potential for success or risk of failure of change initiatives
Organizational change readiness manager	• Assess readiness for change • Develop and implement strategies to increase readiness within organization for change
Organization change adoption lead	• Evaluate effectiveness of change initiative • Develop strategies to ensure adoption of change within organization/engage stakeholders in change
Business change representative	• Work within business to promote change for business • Provide communication link between business and change team and/or project team
Change agents	• Champion change throughout the organization • Be an advocate for change
Stakeholder Support Committee member	• Communicate about change • Share feedback from stakeholders up to change team • Support work of change agents and change team

change initiatives. In addition, he ensured that individuals' service in such roles would be recognized publicly in the organization and be included in evaluation of performance.

Cultural Diversity Matters Here Too!

When organizations call on employees to serve on change teams, as change agents, and on Stakeholder Support Committees to support and push change forward within the organization, they enable for more diverse representation. A challenge in compiling change teams, especially if we consider the ideal size of five to eight members, is ensuring cultural diversity on the team. Consider that many organizations are culturally diverse; even if they only represent one region in the United States. This cultural diversity must be represented on the change team, as well as among change agents and the Stakeholder Support Committee membership. This cultural diversity representation is essential since organizations are comprised of culturally diverse individuals. Enabling and requiring cultural diversity among those focusing on the people-side of change initiatives aids in better engaging a broader group of stakeholders in the change overall.

Providing Training on Change

I cannot recall one change initiative that has not required training of those impacted by the change. Every change initiative will require training in some form. Training should be determined early on during initial change project planning. Table 7.7 provides a list of items that go into determining the type of training necessary to meet the needs of stakeholders.

As is shown in Table 7.7, training may be as simple as a step-by-step document or as complex as a multiday face-to-face training program. The more complex the change initiative, the more training in a variety of formats is required. In such situations, it is imperative to ensure that there is a training manager, an instructional designer, and a training administrator serving on the change team.

While tasks related to training are a component of the project schedule, it is preferred to have these specific tasks overseen by a change manager over the project manager. Training is an essential component to get individuals within the organization to adopt change. A first question on the lips (or in the minds) of individuals who are learning about an upcoming change to be implemented is whether or not they will have the skills to do the job and, if not, how they will get those skills. Training addresses that pressing question.

Table 7.7   Determining training options

Is this the case?	Training options might include...
A minor change to "tweak" a current process	• Updated documentation/"how to"
Multiple process changes throughout the organization	• Updated documentation/"how to" • Step-by-step videos • Observation
Mergers of departments or workgroups	• New skills development • Redefined roles and responsibilities • Teambuilding
Technology upgrades/ deployment of new software	• Classroom sessions • Updated documentation/"how to"
New product or new service launch	• Flyers/handouts ("how to's"—market, sell, support) • Classroom session • Video/virtual session

For a multiprocess improvement initiated and launched in a global organization, I worked with a client to engage a number of stakeholders in shaping and delivering the necessary training across the organization. The responsibilities of these stakeholders were to do the following:

- *Determine training needs*
- *Determine options to deliver the training*
- *Work with Learning & Development and process owners to develop training to build skills and increase knowledge*
- *Pilot test the training prior to full organization-wide roll out*

Figure 7.17 is a template that might be used to capture information from stakeholders to determine training needs.

In addition to understanding the impacted stakeholder's perception of new skills needed or an update to current skills required to work with the change, it is important to understand the stakeholder's manager's perspective also. While the majority of the time these two perceptions are in sync, I have found over the years that on occasion there is a great discrepancy between the stakeholder's perception and his/her manager's perception. This discrepancy must be resolved to ensure the right training is being developed and delivered at the right time to ensure that the stakeholder can effectively work within the changed environment.

Individual/Group Training Needs	
Project: *<name of project>*	Date: *<today's date>*
Stakeholder: *<indiv or group>*	
Impact of change on day-to-day work of stakeholder?	*<low, medium, high>* *Low = minor tasks impacted only* *Medium = at least ½ of tasks impacted by change* *High = more than ½ of tasks impacted by change*
% of job which will change due to this initiative	*<use 0% - 100%>*
New skills or update to current skills needed to perform in role after change implemented?	*Perspective of individual:* / *Perspective of manager:*
Desired training delivery method(s)?	*<virtual, face-to-face, e-learning, document/handout only [document/handout only available for low impact]>*

Figure 7.17   Template to determine training needs

A number of methods that can be used to capture information and gauge training needs, include:

- Surveys
- Focus groups
- One-on-one meetings
- Meetings with senior leadership and other management
- Industry research as well as research external to the industry

A needs analysis conducted by the change team will help to identify the skills gap and determine the training needs that will be necessary in order for impacted stakeholders to successfully work with the change.

Any change project might be considered successful if it finishes within budget and on time, but if the stakeholders are unable to work with the change, the initiative cannot be deemed to be successful. Identifying training needs and developing training programs to meet those needs not only enables stakeholders to work with the change when it is implemented, but also serves to increase their confidence and comfort level with the change initiative. Table 7.8 provides a list of components of a training needs analysis.

Table 7.8   Components of a training needs analysis

Component	Description
Change Impact Analysis (or, Task Analysis)	An analysis of the impact of the change on the stakeholders—specifically, what tasks or work is being impacted by the change. This is aligned to the complexity of the change. This analysis would consider: • Tasks which will continue to need to be performed • New tasks that will need to be performed • Importance of each task as well as complexity of each task • Knowledge needed to perform task (ties into Competencies Required component)
Stakeholder Analysis	An analysis of each individual stakeholder, or stakeholder group, who is impacted by the change. This is aligned to the percentage of change in their job and the impact of the change on their day-to-day work.
Gap Analysis	An analysis of the gap between current skills and knowledge and desired skills and knowledge of each stakeholder.
Competencies Required	A listing of the competencies required for stakeholders to work within the changed environment and be effective in their roles. Examples of competencies include: problem solving, decision making, financial management, project management

The stakeholder analysis component (shown in Table 7.8) can be replaced with the template shown in Figure 7.17.

For smaller, less complex change initiatives (let's assume only one or two departments are impacted), the template in Figure 7.17 can be used as the primary tool to understand training needs. However, for more complex change initiatives, it would be more prudent to conduct a complete training needs analysis as shown in Table 7.8. This information, once captured, might then be summarized at a stakeholder group level to present to senior leadership for approval for training funding.

Figure 7.18 provides a checklist used to ensure that all relevant information is captured *prior* to designing, developing, pilot testing, and rolling out the training.

Do you understand/have knowledge of....	
	The vision for the change. (Consider the vision as it relates to potential training needs, such as increased productivity, improved customer service, improved decision making)
	The types of training options that are available within the organization and have been proven successful within the organization
	The competencies that will be addressed by the training
	Exactly how processes and/or procedures will change and the impact on the stakeholder's day-to-day jobs
	What resources are available, or need to be sourced, for designing, developing, and facilitating the training (e.g., are contractors needed, are there internal subject matter experts, can internal staff conduct the training)
	How job descriptions or the organization chart will change based on the change initiative
	The skills, knowledge, behavior, and competencies required for stakeholders to work within the changed environment
	Which stakeholders are impacted by the change and the level of that impact

Figure 7.18 Checklist to ensure understanding and relevant information needed to develop training

Implementing Change with a Pilot Group

While simpler change initiatives (such as one process being impacted or a change within one workgroup) may be rolled out without pilot testing, it would not be wise to do so with any change initiative that impacts the organization and stakeholders on a larger scale. Pilot testing of change initiatives is a great way to be sure that everything is working as predicted and expected *before* a larger group makes the change.

As shared earlier in this book, the Stakeholder Support Committee could serve as pilot testers to try out a change prior to organization-wide roll out. Pilot testing of larger, complex changes (such as multiple process or procedure changes) will help to:

- Find potential problems in the design of the process
- Determine if training is effective
- Understand challenges in using the new process
- Identify other areas that need improvement

Pilot test groups should be decided upon early on as part of planning the change initiative. As part of this early planning for pilot testing of a change, use the checklist provided in Figure 7.19 to be sure that the pilot group will be effective in testing and evaluating the change.

	Have the following been considered/planned for/created to ensure successful pilot group testing... *Can you answer "Yes" for each statement below?*
	Those individuals selected to participate in the pilot group have the interest and time to participate
	There is a mechanism in place to gather feedback from the pilot testing group
	There is a checklist or other documentation to ensure that the right processes are being tested as part of pilot group testing
	The pilot testing group are individuals who are impacted by the change and have a vested interest in a successful change
	The managers of the individuals selected for pilot group testing support their staff's involvement
	The expectations of pilot group testing participation is clear and concise
	There are clearly defined roles and responsibilities of pilot group testers
	Communications are planned specifically for pilot group testers to ensure they remain "in the loop" during the change project
	There is technology and other necessary structure in place to support pilot group testers
	There is training available, or being created, to ensure the pilot group testers are able to utilize the change
	Time for pilot testing has been scheduled on calendars – even if currently time periods are tentative

Figure 7.19   Checklist for planning pilot group efforts

The checklist provided in Figure 7.19 provides several items to be considered to ensure that pilot group testing is successful. Successful means:

- The way the processes, procedures, or other change being tested will actually be used is what is being tested
- Ensuring that any challenges or barriers that may arise will be discovered during the testing period

If the pilot testing is wrought with problems, fix those issues and then reengage the pilot testers in another round of testing. Don't rush pilot testing! Too many organizations shortchange this process because the change project is running behind schedule. Pilot testing allows stakeholders to *use* the change in a safe environment. It is better that problems are determined before full organization-wide roll out. Once rolled out, it is more difficult to fix issues that arise with a change; and the organization risks overall project failure because the change will not stick over the long term.

It is preferable to include a mix of champions and resisters in pilot group testing. While it is harder to engage and get commitment from resisters sometimes, it enables them to provide feedback and gets them engaged in the change. When their feedback is considered and they see the change is worthwhile, resisters can be converted into champions!

Rolling Out Change Organization-Wide

Organization-wide roll out may be done at once (the entire organization) or in a scaled approach (department by department or in groups)—depending on the complexity and scale of the change initiative, the challenges faced by the pilot testing group, and for a number of other reasons. Table 7.9 provides three options for rolling out change through the organization.

The type of roll out of the change is usually determined early in the initiative so that planning can occur. However, there may be circumstances that will require a change in the type of roll out planned.

One organization determined that roll out would occur by departments. Roll out was determined by impact of the change as well as resources available to implement the change. Early on in change-project planning, the roll out schedule was released as follows: three departments rolling out initially, then another three, and finally the last two. About midway through the change initiative, however, several department heads left the organization to start their own business. Two of these department heads were involved in the initial roll out of the change. It was determined to be too risky to have these two departments roll out as part of the initial wave since these departments were without leadership with longevity. The roll out schedule was altered to reduce risks and alleviate resources that were attempting to adapt to significant changes in their departments.

While there need to be resources dedicated to rolling out the change within the organization, it is of value to get others involved also. For example, after testing, pilot group members can help to facilitate organization-wide roll out of the change by:

- Serving as champions
- Sharing the value of the change
- Helping to work with others who are beginning to implement the change
- Ensuring that documentation about the change is well developed

Table 7.9   Options for rolling out change organization-wide

Option	Description	When to Use...
Sequential	One workgroup, team, or department at a time	• Resources for roll out are limited • Pilot group testing highlighted a number of challenges in roll out • Additional challenges are expected in adapting to the change • Change has a minimal impact on much of the organization • There are more resisters than champions for the change • Each group's needs in roll out are quite different
Group	Multiple workgroups, teams, or departments in stages; for example: • Department A and B first • Department C and D second • Department E, F, G third • Etc.	• While there are a number of champions, there is also a great number of resisters • Roll out needs to be done sooner rather than later • Resources are available to implement the changes but are tasked with other work also • Some groups will benefit from implementing the change sooner than others • Needs in roll out vary from group to group with some consistencies
Simultaneous	Every department, workgroup, and team at once	• There are significant resources available to support roll out • Every department will benefit greatly from the change happening as soon as possible • Roll out needs to happen quickly for a number of reasons • Pilot group testing went very well, minimal challenges encountered • Champions far outnumber resisters • Needs in roll out are fairly consistent from group to group

Pilot group members are a great resource to assist in engaging resisters in adopting the change. These are their peers and therefore, they carry more credibility than a member of management who is demanding that change occurs. Additionally, the change team (these may be many of the committed resources), change agents, and the Stakeholder Support Committee should be involved in the roll out. Change agents and Stakeholder Support Committee members, similar to pilot group testers, can assist in roll out by serving a number of roles, such as:

- Champions
- Support staff
- Trainers
- Sounding boards
- Gatherers of feedback

The team involved in rolling out the change initiative should be engaged early in the change project. This engagement may simply entail understanding that they will serve on the roll-out team and include loosely defined roles and responsibilities and timing of roll out. Certainly, as roll out gets closer, this team will be even more involved in the initiative and should begin working together much more closely to ensure a successful roll out of the initiative. Table 7.10 provides some key responsibilities of the roll-out team.

> In some organizations, the roll-out team may be responsible for developing and facilitating training, documentation, and feedback channels. While, in other organizations, the roll-out team may be responsible solely for coordinating all of these efforts which are managed by the change team. There is no right or wrong way as long as these key tasks are clearly assigned to an individual or a team within the organization.

MAKING ORGANIZATIONAL CHANGE *STICK*

The change initiative can be one of the best planned and managed projects in the organization, but if certain considerations are not made early in the project, it is unlikely the change will stick over the long term. This is a key component that separates one change project from another within the organization and requires significant leadership support and commitment.

Post change implementation reviews enable for ensuring that what the change project was set out to do (its objectives) was accomplished. The group involved in roll out may certainly stay engaged after roll out of the change. For example, in Table 7.10 a key responsibility was measuring against metrics. This effort will continue after implementation, once individuals within the organization have had time to utilize the change. Some changes may be able to be measured fairly quickly—others may need to be in place for a while before they can be measured. Table 7.11 provides a list of activities that are part of post change implementation review sessions.

Post change implementation review sessions should be led by the change leader and the change team. The work of these individuals is not

Table 7.10   Key responsibilities of the roll-out team

Responsibility	What it entails...
Communications	While part of the overall communications plan, these communications will be specifically focused on the roll out and include, for example: pilot group testers, roll-out plan, timing for roll out, feedback channels, etc. The closer to the roll out, the more frequent the communications.
Engagement of pilot group testers	Providing training to pilot group testers, engaging them in the change early on by keeping them in the loop, ensuring time on their calendars to participate in testing, sharing strategies for testing as well as how feedback on the change will be gathered.
Training schedule	Coordinating with those individuals responsible for designing, developing, and facilitating training and collaborating with that group to develop the training schedule for everyone impacted by the change.
Roll-out schedule	Developing the roll-out schedule with support of key leadership, the change manager, project manager, and change team members.
Feedback channels	Developing or facilitating development of feedback channels to ensure data on the roll out of the change to enable for any "tweaks" that need to occur, additional training, etc.
Suggest/determine rewards or incentives to adopt change	Recommend reward or incentive programs to promote and push adoption of the change. Depending on who is on this team, they may be tasked with determining and rolling out reward or incentive programs.
Metrics measurement	Ensuring metrics in place to measure the success of the change and assisting in gathering data to measure against those metrics post-implementation.
Documentation	Assist in or coordinate developing documentation for the change (e.g., how to's).

complete upon implementation of the change. Similar to a project team's work not being completed until lessons learned are captured (last official task of project close out), the change team's work is not done until post change implementation is complete. Some individuals may be involved longer than others and/or some tasks may be transitioned to others.

Some organizations have the change team responsible for initial data collections as they relate to measuring against metrics; but then transitions that task on to management of the impacted department or workgroup.

Table 7.11   Activities of post change management review sessions

Activity	Review...
Charter and scope analysis	Ensure that the change closely matches what was supposed to be accomplished per the Charter and Project Scope
Documentation	Ensure that documentation clearly describes the change and provides sufficient detail for those who will need to use the change
Feedback/Employee satisfaction	Ensure feedback channels in place and efforts are made to engage individuals using the change in providing feedback about the change
Metrics measurement	Develop a plan to gather data/data collection to measure against metrics to ensure the change is working as desired and expected
Training	Ensure that additional training options—in a variety of forms—are available to those who need further training
Lessons learned	Work with the project team to capture lessons learned, with a specific focus on lessons learned in engaging people in change
Additional needs	Determine if, or capture information around, additional changes that may be necessary to support or enhance the current change

Organizational change can only last when efforts are made to ensure it lasts. Change doesn't stick just because it was implemented, it sticks when there is a concerted effort to engage individuals in the change early on and throughout; and ensuring that these individuals are engaged after the change is implemented.

Deploying the Change Agents

The work of the change agents is not done when the change is implemented. Deploy change agents—as well as Stakeholder Support Committee members—to continue to champion the change. These individuals can assist in gathering feedback, providing *just-in-time* training, and in capturing lessons learned.

> For a pharmaceutical client, each time a change is implemented, a select number of change agents are selected to serve in the role of moving from department to department to ensure that employees have what they need to work with the change. They may provide training, gather feedback, answer questions, or just encourage the employees to give the change a try. This is a coveted role in

the organization, and those individuals selected are provided time away from their daily workload to serve in this capacity. Individuals are selected through nomination by the change manager, change leader, and their own manager—based on their work in the change agent role.

Don't just deploy the change agents without a plan! Be sure that each change agent has:

- Received training in how to manage individuals who may be upset or frustrated with the change
- A list of specific individuals or departments with whom they will interact
- A schedule (developed in collaboration with each change agent) for interactions with those working with the change
- A forum and process for reporting back on their efforts
- Support from his or her manager so that they can be effective in their roles
- A way to collaborate and share information/feedback among the team

The number of change agents or Stakeholder Support Committee members deployed after implementation will be dependent on a number of factors, including:

- Number of champions versus resisters
- Complexity of the change
- Scope of the change
- Urgency in getting individuals to adopt/utilize the change

Checking in with People

In addition to having feedback channels in place to gather data about what is working or not working with the change, it is essential to utilize a number of other methods to check in with those using the change. Options for checking in with employees who are working with the change include:

- Collaboration sites
- Surveys
- Department meetings
- Focus group sessions
- One-on-one meetings/observation

- Sharing information of other employees who are working success-fully with the change

  One change agent would always share with individuals whom she was checking in on about how she struggled to adopt change in her previous organization. She would share that she doesn't always like change herself since she is comfortable with the status quo. She would then share ideas regarding how she would eventually come around to accepting and embracing the change. This change agent did a great job in engaging others toward accepting the change because she shared her own struggles.

By providing a variety of ways for employees to provide feedback on the change, the change team is more likely to gather the information they need to ensure the change is successful and to continue to engage people in using the change.

As a best practice, make sure that change agents and others deployed to check in with employees are reporting back on their findings in a consistent way.

Continuous Evaluation and Improvement

The more complex the change initiative, the more time that must be spent considering its adoption and whether or not further improvements are necessary.

For one organization, while it was thought that every area impacted by the change was considered early on, it was found—upon full roll out—that a few other processes needed to change to support the change that was just launched. Because the other processes were fairly minor, they were missed in early evaluations and in pilot group testing. However, during organization-wide roll out, those individuals who used the minor processes noticed the problems. They brought those issues to the attention of the change team and changes were made to accommodate the employees.

It is important to remember that the efforts of a post-implementation process will promote continuous improvement of future change initiatives. There needs to be a process and system for capturing and sharing this data to initiate continuous improvement.

Once a change initiative is launched, it must be evaluated to ensure it is actually working for everyone who needs to work with the change and that it is not negatively impacting any other areas of the business. Evaluation of change initiatives can be a bit more challenging to manage with a timeline. Unfortunately, there is no magic number concerning when the change team and others engaged in post-implementation work might stop their efforts in evaluating the adoption of change.

Any of the following will require more time spent in post-implementation to ensure that people eventually adopt the change:

- Complex changes
- Significant resisters to the change
- Past change initiatives that have failed
- Change initiatives that have been wrought with problems throughout their life cycle
- A lack of sufficient communications early on and throughout the change initiative
- A lack of reward or incentive structures in place to enforce the use of the change

Organizations can increase the adoption of change through post-implementation efforts by doing any or all of the following:

- Ensuring regular discussions around the vision for the change and the benefit of the change *for the individual*
- Ensuring there is a definitive deadline to adopt the change (this is particularly important for process and/or technology changes)
- Asking those who have adopted the change to talk about the benefit and value they have found in working with the change
- Asking those who are struggling with the change what else might be done or what else they need to help them adopt the change
- Offering (and promoting) rewards for those who adopt the change
- Ensuring sufficient and various opportunities and channels for individuals to be trained, coached, or get to support in adopting the change

Once it has been determined that enough individuals have adopted the change, the change is successfully being utilized within the organization, and no further issues have arisen in working with the change, the evaluation work of the change team is finished.

Remember that some very complex, organization-wide changes may take a significant amount of time to fully evaluate the success of the change. Certainly, changes that are implemented in stages, rather than all at once, will require more investment in monitoring and evaluation.

Data captured from change initiative post-implementation efforts will allow for a better prediction of how long the change team will need to invest in monitoring to be sure a change is adopted. However, people are unpredictable and, effectively, a wild card when it comes to change. The fact that a past change was adopted quickly is not an indication that a future change will be adopted quickly. This does not mean to imply that regularly evaluating each change and striving for continuous improvement through lessons learned is a waste of time. It certainly is not! Much can be learned about how to manage the next change initiative—maybe even better than the last one. More information will be shared about effectively capturing and applying lessons learned in the next chapter.

This book has free material available for download from the
Web Added Value™ resource center at *www.jrosspub.com*

8

CONTINUOUS COMMUNICATION AND ENGAGEMENT IN CHANGE

> "To change ourselves effectively, we first
> have to change our perceptions."
> Stephen R. Covey, *The 7 Habits of Highly Effective People: Powerful Lessons in Personal Change*

Throughout this book so far, we have discussed the necessity of continuous communication and constant engagement of employees in each change initiative that is launched. Certainly, organizations that have processes and best practices for engaging and communicating with employees during change do a better job of implementing a successful change overall.

However, the ability to engage employees in change *even when a change is not being implemented*, is what sets the best organizations and the strongest leaders out from the pack. Leaders who consider change a strategic component of the organization and the primary reason for launching strategic projects will understand and appreciate the value of continuous communication and engagement of employees outside of conversations around one single change initiative.

As organizations consider change part of the normal way of doing business, and enable for an organizational culture and structure that supports that *norm*, then—and only then—can change management be considered a strategic component within the organization.

In order to get employees more interested and engaged in change, one Chief Executive Officer (CEO) launched an initiative that rewarded individual contributors and supervisory level employees as follows:

- *$5 fast food or beverage gift card for each idea offered that would initiate improvement in any of the following areas: manufacturing processes, customer service, operations, or sales*
- *$50 debit gift card if the idea is selected as one of the strategic projects to be launched*
- *$500 bonus if the individual who offered up the idea is willing to play a key role in helping to champion the idea throughout the organization for the duration of the project*
- *$1,500 bonus if the idea, once implemented, is adopted by 80% of the employees within one year of roll out*

The CEO has found that this initiative accomplished the following in one year:

- *75 ideas offered for improvement*
- *20 of the 75 ideas accepted as strategic projects (over a 3-year time period)*
- *All 20 of the strategic projects were adopted by at least 80%— with one garnering 92% adoption*

This initiative was continued for another two years.

In this chapter, the focus will be on how to move an organization toward seeing change as the norm through establishing processes, procedures, best practices, and technology to support and encourage regular communication and engagement around change.

LEARNING LESSONS FROM PAST CHANGE INITIATIVES

Lessons learned are knowledge gained by past experiences that will have a positive impact on an organization and its employees. Lessons learned could be lessons from something that has gone well, or something that was a failure or a negative experience. Lessons learned, when applied over time, enable for significant improvement in how projects are managed, how change is adopted, and how the work is done overall. Lessons learned also encourage communications and collaboration around how to improve—which leads to change.

A part of every change project launched in the organization should be capturing and applying lessons learned.

For every change project launched, regardless of whether the initiative is simple or complex, the first activity is to look at past change initiatives to see what lessons learned can be applied to the current initiative. The last activity is a meeting to review lessons learned that were captured during that change initiative. Both of these crucial meetings should encourage conversations around change— what's working and what's not working with change initiatives.

Figure 8.1 provides a process for capturing and disseminating lessons learned from every change initiative launched within the organization.

Figure 8.1 is just one possible process or framework for capturing and sharing lessons learned. In general, there must be a process in place that is used to:

- Collect information about change initiatives and in particular, how effectively employees were engaged in change
- Ensure that the information captured is applicable within the organization, is valuable to improve future change initiatives, and is complete and accurately gathered
- Ensure that the information is stored in a repository
- Ensure that information in the repository is distributed throughout the organization and used to improve engagement in the next change initiative that is launched in the organization

Getting employees involved in this process keeps them engaged in change.

Technology increases the possibilities for improved collaboration and sharing of information about change initiatives. The checklist in Figure 8.2 provides several considerations when deciding on a tool to use to capture and share information on change management initiatives.

A tool selected by one organization, unfortunately, was very limited. Lessons learned data could not be categorized easily, and metadata was limited to author, date, and location. There was no ability to customize dashboards. The tool was selected when the company was in its infancy and as the organization grew, was no longer valuable. Unfortunately, by the time the organization selected another tool that was more adaptable to their needs, lessons learned had been collected sporadically on paper. There was literally a file cabinet full of lessons learned that needed to be entered into the new tool! Prior to this work being completed, change managers rarely, if ever, kicked off the change team with a review of lessons learned on

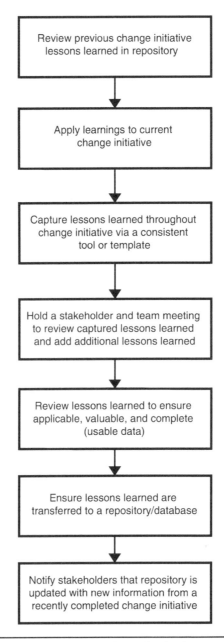

Figure 8.1   A process for capturing and disseminating lessons learned around change

	The tool is compatible with other technology in use within the organization
	The tool can be accessed by every employee in the organization
	The tool is easy to use
	Documentation, "how to's" and other training materials are available
	The tool supports the processes around capturing lessons learned
	The tool enables for easy searching of data and information
	The tool is customizable to the organization
	The tool enables for collection of metadata
	Personalized & customized dashboards with communication widgets are available

Figure 8.2   Checklist: evaluating tools for capturing and sharing information about change initiatives

past change initiatives. Thus, the same mistakes kept recurring. As the Chief Financial Officer (CFO) put it, "We just kept reinventing the wheel. It was so frustrating!"

While lessons learned should be captured for *every* type of project implemented in the organization, here the focus will be on the specifics of capturing lessons learned for change projects, with attention paid to how well the change team managed people's expectations around change, maintained communications around change, and kept them engaged in the change.

Certainly lessons learned around change management may be combined with lessons learned around project management; but it may be easier to separate the two lessons learned meetings and then review the data gathered from both meetings. This often keeps the groups participating in the meetings smaller and easier to manage. Additionally, smaller groups usually produce more in-depth conversations.

Lessons learned will generate communication about a change initiative and specifically what went well during the launching, implementing, and rolling out of that change; along with where things must be done differently to enable for a more successful change initiative the next time around. Figure 8.3 provides a template to capture lessons learned from every change initiative. This template is specific to capturing lessons learned around change management practices.

This template to capture lessons learned will help to gather information about what went well in managing the change, as well as where

Change Project: *<name>*	Change Manager: *<name>*	Date of Lessons Learned Meeting: *<date>*
Change team, change agents and Stakeholder Support Committee members in attendance:	*<names and roles>*	
Lessons Learned from Negative Impact Situations		

Area, Process or Component reviewed	Issues that Arose	Actions Taken to Correct and When	What Worked Well?	Where do Improvements Exist?	What will be done specifically to improve?
<Include area, process or component that was the source of a negative impact that was reviewed in the lessons learned meeting>	*<Be specific about issues that arose that may have impacted the project in a negative way>*	*<Delineate what, if any, actions were taken and when they were taken to correct the situation.>*	*<Be specific about what went well with actions taken.>*	*<Be specific about what could have been done to avoid the issue or to improve upon actions taken.>*	*<Be specific about steps that will be taken. Team might refer to another document for details on steps for improvement.>*

Lessons Learned from Positive Impact Situations		
Area, Process or Component reviewed	**Positive Impact Realized**	**How Will This Positive Impact be Shared?**
<Include area, process or component that was the source of a positive impact that was reviewed in the lessons learned meeting>	*<Be specific about what positive impact was realized that was of value and benefited the change project.>*	*<Be specific about what will be done to be sure that this information is shared among all employees and used from project to project.>*
What additional information was shared or ideas gathered from this lessons learned meeting:	• *<Consider here: how effectively employees were engaged in the change, how resisters were converted to champions, how a broader group may have been involved in engaging across the organization, the effectiveness of the various groups involved in managing people's expectations around change, etc.>*	
	• *<Delineate additional information in separate rows for easier scanning and review of information gathered.>*	

Figure 8.3   Lessons learned template

improvements could be made to ensure a better process and better engagement of employees the next time around. Use categories such as *communication on change, engagement of employees, the vision for change, training on the change*, etc., in order to ensure that lessons learned are tied back to what is important for a change initiative to be successful.

As a best practice, at the start of a change management lessons learned meeting, discussions should focus on areas of improvement only. Names of those who may have caused issues or been involved in an issue are just not necessary. However, when things have gone well, the team should highlight those individuals who contributed positively. This inspires increased participation in the discussion and increases the comfort levels of those involved. People should be reminded that everyone, at some point, makes mistakes and while it is great to use them as opportunities to learn and improve, there is no need to harp on the mistakes. During the meeting, a "to do" list is maintained and members of the change team, change agents, Stakeholder Support Committee members and other participants often take on smaller initiatives with a focus on improving change management within the organization. These mini projects effectively keep a focus on change within the organization and enable for continued communications and engagement on best practices around change management.

Figure 8.4 provides a partially completed template from a lessons learned meeting.

Just capturing information in the template is not sufficient. Something must be done with that information if it is to be useful in the organization. There needs to be a forum to share this information as well as a process for following up on decisions that are made from the lessons learned meeting. Internal sites, databases, or knowledge bases are tools to store information gathered from lessons learned. As a best practice, ensure that whatever tool is used, it is easily accessible by *everyone* in the organization, *regardless* of their location, and makes it easy to search for and find information.

I use Microsoft SharePoint™ sites to share information gathered from lessons learned from change management initiatives. The site is accessible by all employees, regardless of their role. It includes information about past and upcoming change initiatives, serves as a way to acknowledge individuals who have served on change teams, as change agents, and in Stakeholder Support Committee roles. It allows searching by a number of keywords and project types as well as project names in order to learn more about projects and lessons learned from those projects. Not only is this site used by those working on change initiatives, but over the years, many employees have started to use the site to suggest changes as well as to engage in conversations that have led to a number of change initiatives being launched by leadership.

Change Project: Implementation of CRM System	Change Manager: Alison		Date of Lessons Learned Meeting: April 5
Change team, change agents and Stakeholder Support Committee members in attendance:	Change team members: John, Amanda, Sarah, Julian, Lucas, Rafael (in person) Change agents: Bill, Debra, Fay, Donald, Lisa (in person); Sara, Thomas, Anna, Roderick (virtual) Stakeholder Committee members: Paul, Julianne, Ann, Steven (virtual)		

Lessons Learned from Negative Impact Situations					
Area, Process or Component reviewed	Issues that Arose	Actions Taken to Correct and When	What Worked Well?	Where do Improvements Exist?	What will be done specifically to improve?
Early communications about change	• Insufficient channels to get the word out about the change • Employees felt disconnected, left out and were upset and frustrated which impacted work getting done	• Additional channels added after 3 months • One-on-one and small group meetings held after 3 months to address concerns of those who felt channels of communication were insufficient • Email sent from Change Leader apologizing for lack of communications	• Additional channels, once launched, enabled for addressing concerns and improved communications for the balance of the project • One-on-one and small group meetings enabled for managing individuals who felt left out of the process.	• Ensuring a variety of channels decided upon and utilized prior to launching change initiatives • Addressing issues, such as this one, much earlier on. By addressing after 3 months, employees got angrier over time and felt ignored.	A committee will be tasked with developing a list of acceptable channels for communicating on change initiatives to include pros and cons of each channel with guidelines for selecting channels for use on change initiatives. Committee work will begin on April 8 and end by May 8 with a report due to senior leadership on May 12. Final selection of channels will be made by no later than May 20.

Lessons Learned from Positive Impact Situations		
Area, Process or Component reviewed	Positive Impact Realized	How Will This Positive Impact be Shared?
Use of Stakeholder Support Committee	Stakeholder Support Committee, which included representatives of each division, department and geographic location, provided specific feedback to the change team that enabled for making adjustments on communications about the project as well as training needs that increased the overall project success.	The value that the Stakeholder Support Committee provided will be shared with the entire leadership team. Stakeholder Support Committee members have agreed to develop a handbook as well as best practices for future committee members.
What additional information was shared or ideas gathered from this lessons learned meeting:	• Resisters may have been converted more easily and sooner had the change team recognized early on the number of resisters that existed. • Clearer roles and responsibilities for change agents and Stakeholder Support Committee members should be defined for future projects to ensure better use of limited resources' time	

Figure 8.4   Partially completed example: lessons learned

Capturing and applying lessons learned from change projects will encourage improved communications on change overall. Additionally, it often adds yet another change project that employees might undertake. In Figure 8.4, one outcome of the lessons learned meeting was to form a committee and to task the committee with developing a list of acceptable channels for communicating on change initiatives. This is, effectively, the beginning of another change initiative. Once a final selection of approved channels is decided upon, the change initiative will entail developing or securing tools for communication and collaboration, implementing those tools within the organization, and training individuals on the use of those tools.

Utilizing lessons learned processes and best practices on change initiatives enables continuing the conversation around change within the organization. It shows the employees that change is a norm in the organization and that there is a desire to ensure that change is always done well.

Look back at the story shared earlier in this chapter about the CEO who got employees more interested and engaged in change through offering incentives to put forward ideas and incentives to work on change initiatives. This idea came about through a lesson learned from a change initiative where employees who served on a Stakeholder Support Committee felt that there was limited appreciation for their efforts serving on the committee. After the initiative ended, the CEO still acknowledged employees' participation in change initiatives, but without a monetary incentive. At this point, the monetary incentive was not necessary—change was the norm in the organization. Table 8.1 provides several potential methods to collect data about lessons learned.

These options help people understand the successes and challenges of launching and implementing change initiatives. Regardless of which option is used to collect data about lessons learned, it is essential that the following questions are being answered in the collection:

- What went well with the change initiative?
- What did not go well with the change initiative?
- What might be done differently to increase adoption of the change?
- For problems that were not solved during the change initiative, what measures can be taken to solve them the next time or prevent them altogether?
- What new best practices from this change initiative should be implemented in future change initiatives?
- What else should be shared to enable change initiatives to be improved upon?

Table 8.1   Methods to collect data on lessons learned

Method	Best Uses
Surveys	• To gather information from a large number of individuals • Can be completed anonymously if necessary • Can easily compare and analyze data gathered • To gather statistical information/quantitative data Surveys can include open-ended, comment-type questions and the ones launched by the author often do. This is a great way to gather information from large numbers of individuals but does require more effort in analysis.
One-on-One interviews	• Great way to follow up on results from surveys, focus groups, or observation • Can be used to gather detailed information specific to individuals • Provides qualitative data When groups are small, the author will frequently conduct one-on-one interviews which enables better conversations about the change initiative and enables more robust collection of lessons learned data.
Observation	• Enables understanding exactly how individuals are performing/using the change The author uses observation in particular for process improvement change initiatives. Observation enables viewing how the employees are *actually* using a process and not just how the process was documented. Observation works great to understand how effective the training for a change was for those impacted.
Focus groups	• Enables reaching out to a broad audience (such as an entire department or division) • Can be used to build on ideas, thoughts, suggestions, and comments of others When holding focus groups, make sure a variety of options (time/date/location) for attending are made available and that a seasoned facilitator is used to be sure that the room is well managed and the right information is collected.

Lessons learned should be captured on change initiatives throughout implementation of the change project—from project kick off through to project closure, with a final meeting at the end of the project to discuss the lessons learned that were captured. Figure 8.5 provides an example of a process that a health and wellness organization used to capture lessons learned early on and throughout change initiatives for review at the end of projects.

This particular organization has recently hired a lessons learned coordinator who is responsible for disseminating lessons learned throughout

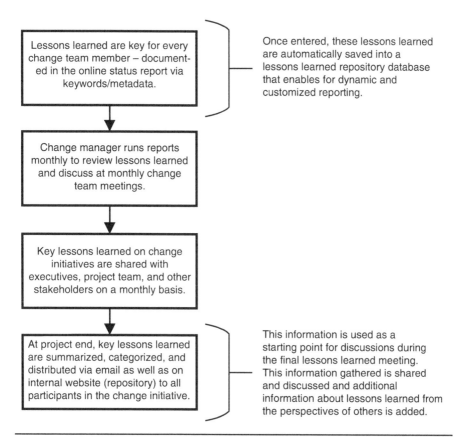

Figure 8.5   A process for capturing lessons learned early on and throughout the change project

the organization, maintaining the repository, and working with each change team for every change initiative launched, in order to identify lessons learned that will be of the most value to the team on their particular project. This same client will be hiring a lessons learned leader within the next six to nine months. The role of this individual will be to work with each change team on every change initiative, in order to ensure that lessons learned are captured in a consistent and usable format and with the goal of ensuring continuous improvement and engagement in change from throughout the organization. This individual will be responsible for overseeing the repository and leading lessons learned meetings.

THE VALUE OF MAKING CHANGE THE *NORM*

There are three types of organizations that implement change:

1. Those organizations that change *on occasion* just so that they can maintain the status quo: In this organization, an example of change may be to update technology because the old technology just won't support the software programs in use and in order to continue using that software, the technology must be updated. (Change is not the norm here.)
2. Those organizations that change because their proverbial backs are up against the wall: In this organization, customers are being lost to the competition and revenues are decreasing. The organization changes because they are forced to do so. (Change is definitely not the norm here either!)
3. Those organizations that embrace and encourage change: They hire and nurture leaders and employees who champion change and look for ways to continue to change and adapt in order to remain competitive. These are the organizations that will be the focus of this chapter and the next chapter. (Here, change is the norm!)

When change is the *norm* within an organization, it is less worrisome for employees—it becomes expected. Employees are more likely to embrace change and see it from a positive perspective. When change is the norm, it is usually better managed overall. It is not chaotic, but rather is planned for and employees adapt to change much more quickly. Leaders encourage employees to regularly change; they support conversations around change and provide opportunities for employees to try something new. They provide resources and funding for change. They understand that risks exist and enable for calculated risk-taking. Capturing lessons learned is important in these organizations, as is sharing those captured lessons learned.

Lessons learned enable an organization to move toward being one where change is the norm. A commitment to lessons learned—capturing, sharing, and utilizing—shows a commitment to improving how change is launched, implemented, and managed.

Refer back to the example in the previous section about the health and wellness organization and their process for capturing and disseminating lessons learned (Figure 8.5). Through the hiring of a lessons learned coordinator, the goal of hiring a lessons learned leader, and the development and implementation of a repository specifically for lessons learned from change initiatives, the organization has shown, and continues to show, a commitment to supporting change within the organization.

For one healthcare organization where, employees will note, change happens *almost weekly*, the organization has over the years empowered the employees to take ownership of change. This has been done through seeking out recommendations by employees for areas of change. As suggestions on how to improve through change initiatives became more prevalent in the organization, the Chief Human Resource Officer incorporated change management into the performance review process. Each individual in the organization, regardless of their role or responsibilities, was tasked with identifying and implementing a minimum of two personal work-related change initiatives each year. Over the last two years, senior leadership has seen an increase in efficiencies in work getting done with an increasing number of initiatives coming in on time and within budget and an increase in customer satisfaction ratings overall. This is a great example of change being the *norm* in an organization. When organizations make change the norm, they realize a number of benefits, as shown in Table 8.2.

One change initiative may enable an organization to realize some of these benefits. However, the value of the benefits increases exponentially with each and every successful change initiative launched in the organization. When change is the norm, it becomes easier to get change initiatives implemented, get individuals involved in working on the change, and adoption of change increases among employees. In organizations where change is the norm, change simply becomes a part of the everyday work of the organization.

An organizational change management plan, similar to an organization's strategic plan, provides a tool for furthering the organization's commitment to change as a norm. Such a plan—if looked at strategically—should include, but not necessarily be limited to:

* A vision for change as the norm

Table 8.2   Benefits of making organizational change the norm

• Easier integration and adoption of new technologies to improve the business	• Increased retention of employees through regular learning of new skills
• A regular focus on the needs of the customer and therefore the ability to adapt to meet customer needs	• Problems become an opportunity rather than a challenge
• Ability to manage through a downturn in the economy	• Improved product and service offerings to customers
• Continuous growth of the organization and its employees	• A competitive advantage

- A listing of resources committed to change—such as change leaders, change managers, change agents, lessons learned coordinators, etc.
- Best practices, processes, and procedures on initiating and implementing change
- Communication channels devoted to gathering feedback on change
- Communications channels devoted to engaging the organization in conversations around change
- The availability of budget monies allocated to innovation and growth programs within the organization
- Technology used to support conversations around change

KEEPING THE CHANGE CONVERSATION GOING

Keeping the conversation around change going can be quite simple. For example, an organization that utilizes *suggestion boxes* is engaging employees in conversations about change. A suggestion leads to thinking about change. However, the organization only keeps the conversation going when they actually *do something* with the suggestions offered up by employees. This is certainly one simple way to engage employees in conversations around change. In addition to utilizing suggestion boxes, Table 8.3 provides a few additional ideas to keep the conversation around change going.

Knowledge Management and Collaboration Best Practices

Continuous communication around change must be enabled in order for any change to happen. Technology is an effective way to prepare for it. Certainly supporting conversations through formal and informal channels—such as one-on-one meetings, all-staff meetings, through lunchtime get-togethers, or through leadership informally asking employees how things might improve while standing in front of the coffee machine are all methods to encourage conversations. But technology can support those

Table 8.3 Ways to keep the change conversation going

• Incorporating change conversations in all-staff or all-hands meetings	• Incorporating change conversations into team or department meetings
• Utilizing change discussion groups on internal websites or portals	• Creating roles for change conversationalists and deploying them to initiate conversations about change
• Hosting breakfast or lunch change discussions sessions (and provide breakfast or lunch!)	• Initiating a monthly Change Newsletter that focuses only on positive change within the organization

conversations and ensure they continue. Through the use of technology, organizations can initiate conversations around change, encourage ideas for improvement to be submitted, provide tracking of ideas submitted, as well as sharing best practices. There is significant knowledge and best practices involving change that might be lost in an organization as individuals come and go in their jobs. Some employees may just not know how to share their ideas. Technology can ensure that knowledge, ideas, and best practices are retained.

> *Microsoft SharePoint® helps all employees engage in conversations around change. The tool is used to:*
> - *Start conversations around best practices*
> - *Provide a forum for sharing ideas*
> - *Submit suggestions to improve processes and workflow*
> - *Share articles and white papers related to change*
> - *Maintain lessons learned information and communications plans from previous change initiatives*
> - *Provide templates for change management*
> - *Provide a list of internal resources who are considered—by their peers—to be change agents*

Knowledge management and collaboration go hand-in-hand. With knowledge management:

- Employee productivity increases,
- Problems are solved more rapidly,
- Collaboration increases,
- Ideas are shared across the organization,
- Changes happen more frequently,
- Employee knowledge and skills increase, and
- There is a broader understanding among employees as to how they fit into the organization and how they can contribute beyond their own workgroup, team, or department.

Collaboration tools—such as Microsoft SharePoint® discussed in the example above—enable for collecting and leveraging the knowledge within the organization. Given the global nature of organizations today, simply utilizing face-to-face meetings to share information and best practices is hardly sufficient. Technology is necessary, especially in an environment where employees often work remotely.

A colleague of mine works in a national organization that uses Yammer to collaborate and communicate across the organization regarding change initiatives. Of the nearly 1,500 employees who comprise the organization, many of them work remotely from their home offices. When change initiatives were launched in the past, these employees were often forgotten. They lacked an understanding of the vision for the change in the best case scenarios. In the worst case scenarios, older technology would be replaced and employees would not even know about it. Through the use of Yammer, each employee in the organization is connected with what is happening. They learn of change initiatives early on and are encouraged to participate through serving on virtual teams to help implement and test the change. This tool is also used to allow ideas for change initiatives to come from all employees and be voted on as to whether or not to implement those ideas on a quarterly basis.

For organizations that do not currently capture and share knowledge, moving toward being one that does is a change initiative in and of itself. Best practices around knowledge management are essential to initiate regular communication and engagement in change. Figure 8.6 provides a process for launching a knowledge management practice within the organization to enable for increased collaboration and sharing of best practices, which will enable for increased conversations around change.

Figure 8.6 provides a step-by-step process to launch organizational and culture change efforts for implementation of a knowledge management practice in the organization.

When launching a change initiative to implement a knowledge management practice, be careful of how the change is framed for employees. As with any change—share the vision. For example, the vision for a knowledge management practice may be to enrich the organization and its employees through sharing knowledge and increasing collaboration. A friend of mine was working in an organization that wanted to launch a knowledge management practice. In sharing the reason for the practice, my friend's boss told him and others during a department meeting that due to the recent turnover in the organization there was a fear that knowledge was being lost that needed to be captured. This may well be true; but not the best way to sell a change. Needless to say, that statement was not well-received and quickly spread through the organization. It eventually was being shared from employee to employee that the

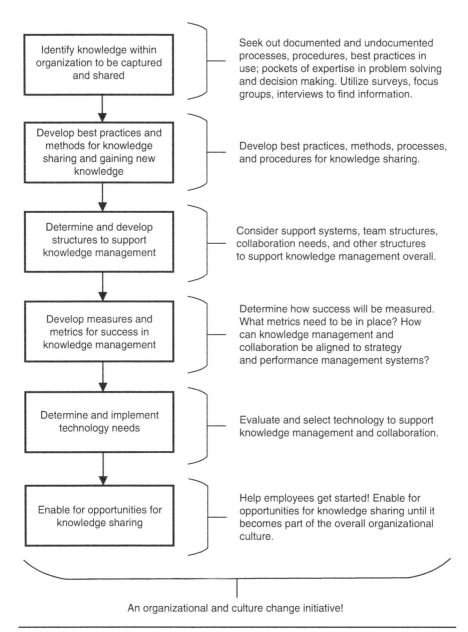

Figure 8.6   A process for launching a knowledge management practice

organization would likely do layoffs and was trying to learn what everyone knew so they could reduce the headcount. When my friend asked another leader for clarification on the initiative to launch a knowledge management practice, he learned that the organization wanted to create improved collaboration and reduce time spent trying to figure out problems that, often, had already been solved elsewhere in the organization. My friend was told, also, that employees had much to share and there was no good infrastructure in place to enable sharing. Additionally, a few employees had noted to their management that they would love a forum that allowed for sharing best practices and collaborating to solve problems. While certainly a concern may be that knowledge is being lost and must be captured and retained; such a statement only serves to disengage people from the change and make them want to hold even tighter to their knowledge to save jobs that they perceive may be lost. This will cause the change initiative to fail.

As a best practice, do some research prior to implementing a knowledge management practice. It is highly likely that there are already pockets of knowledge being captured and shared within the organization. Once these pockets are found, talk with employees who are sharing knowledge already to find out:

- What's working for them?
- What is valuable to them in sharing knowledge with others?
- What would enable them to continue to share knowledge?
- How can leadership support their efforts in sharing knowledge?

Use this information to frame a vision for implementing a knowledge management practice and ask those employees who are already sharing knowledge to help in championing and supporting the effort.

One retail organization encouraged knowledge sharing among employees by tasking each employee to share at least one best practice outside their own workgroup. This goal was included in their annual performance goals. For those employees who went above and beyond and regularly shared best practices and knowledge with others—such as by mentoring, assisting on initiatives outside their own domain, etc.—they would be rewarded with a bonus at year's end as well as be highlighted in the company's quarterly newsletter. The more knowledge sharing that occurred, the more the amount *of the bonus earned. One employee, who mentored five new hires and regularly shared his best practice by creating and sending a newsletter entitled "Getting the Work Done More Easily" received a bonus of $2,500 at the end of one year as well as a personal written "thank you" from the CEO. Once word spread about the bonus and the personal "thank you" from the CEO, a number of other employees jumped on the bandwagon and began collaborating more frequently across the organization and contributing to the "Getting the Work Done More Easily" newsletter. Within a couple of years, more than 50% of the employees were writing articles and tips for the newsletter which was sent out on a weekly basis. The organization leadership had succeeded in changing the culture to be one of sharing and collaboration. All without launching a formal change initiative to do so!*

MAKING CHANGE CONVERSATIONS A PART OF ONBOARDING FOR NEW EMPLOYEES

Organizations can engage employees in regular change conversations from the minute they walk in the door, as part of an onboarding program.

An entertainment client challenges each new employee on day one to evaluate how the work is done in their area. If they can find a way to improve the method in which work gets done within 12 months of working in the role, they will receive a bonus equal to 20% of their annual salary. Additional requirements of the challenge include a positive impact to the team, department, or external customer through, for example, shorter time-to-market to get products out the door, reduced costs, increased revenues, or improved customer service.

Table 8.4 provides ideas to enable and encourage continuous communication and engagement of new employees in change.

These ideas assist in getting employees comfortable with a culture of change and promote the idea that change is valued and embraced in the organization. New employees in particular bring new experiences to the organization. While leaders want employees to adapt to the culture of the organization, when that culture is one of regular communications and engagement in conversations of change, it is essential to let employees know that their ideas to improve are welcome.

Table 8.4 Ideas to encourage conversations on change for new employees

• Provide guidelines on areas "ripe" for change, for example—work processes, customer service	• Provide tools to encourage collaboration on change and train new employees on those tools
• Challenge employees to change how work gets done to improve workflow	• Enlist new employees to help out on change teams
• Ask employees questions about previous experiences	• Share stories of change initiated by employees to encourage new employees to offer their ideas
• Check in with employees after they have gotten to know their role—what ideas do they have to improve the workflow?	• Provide mentors who are champions of change and encourage and facilitate conversations around change

Figure 8.7 is a list of some best practices and tips for employees who want to bring change to their new role.

When encouraging new employees to consider engaging with others in conversations around change, it is important to reiterate the need to build relationships and get to know the job *first*. It is impossible to successfully share best practices and knowledge without *first* building relationships, listening to others, and truly understanding where improvements might exist.

A software development firm assigns new employees to a change team task force. The task force is an ongoing initiative within the organization. Employees from throughout the organization are challenged to review certain processes and procedures and determine changes that might be made to accomplish any of the following long-term goals:

- *Reduce expenses*
- *Increase revenue*
- *Improve customer support*
- *Improve profitability*
- *Enable for efficiencies in getting software out to the market*

The new employee who joins the team is asked to actively participate by bringing their own experiences from past organizations to conversations. Effectively, the organization asks new employees to bring an outsider's point of view to the conversations around change. The organization is not expecting that the new employees

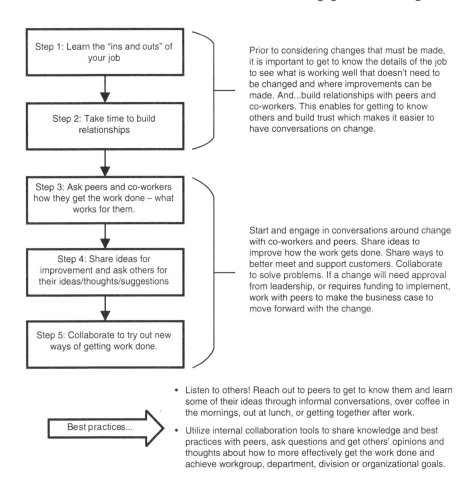

Figure 8.7   Steps and tips for new employees to aid in introducing change

will be very involved in the change initiative overall since they are just learning the job themselves, but rather this component of on-boarding encourages the new employee to build relationships with peers and become comfortable thinking about change.

9

CONTINUOUS—BUT NOT CHAOTIC!—CHANGE

"Look around you. Everything changes. Everything on
this earth is in a continuous state of evolving, refining, improving, adapting, enhancing…changing. You
were not put on this earth to remain stagnant."
Steve Maraboli, *Life, the Truth, and Being Free*

Continuous change, that is *not* chaotic, can only happen in instances when
there is trust between employees and leaders. When trust does not exist, every
initiative launched in the organization is questionable and met with resistance.
Continuous change can only happen when it is tied to solving a problem or
accomplishing an objective, and it makes sense with everything else going on
in the organization. When it is not tied to a vision, change doesn't make sense;
employees can't figure out why the change is happening. When change is chaotic—or appears to be so—employees suffer from fatigue and become disengaged and disheartened; it seems that the organization is always changing and
employees are pulled in too many directions at once.

Continuous change can only happen when employees are encouraged
and supported to always improve on how work gets done. They need the
skills, knowledge, time, and tools to be able to change. When employees
are specifically asked by leadership, "*How can we help you to change?*" the
perception among employees is that leadership understands the challenges
associated with change and are willing to support that change. This inspires employees to see change in a positive way rather than negatively.

A colleague of mine shared this story:

> *She was supervising a team of ten employees who were tasked with implementing a change initiative within the department. Her manager, who launched the initiative, told her and her team that the initiative was the most urgent project the department would be working on and would be the only focus of the group over the next six months. While the initiative was definitely important, she shared with me that this was the third time in the last nine months that he had told the group that an initiative he launched "was the most urgent one." It had become a bit of a joke among the group. Every initiative seemed to be the most urgent one to this manager. The team walked away shaking their heads at yet another initiative that would take them away from their jobs and end up being a waste of their time.*

The nature of customers in today's competitive marketplace makes the need to continuously adapt and change a necessity and an urgency. Customers are demanding. They want excellent service, improved products and services, and access to the company's employees for technical support. They want their needs to be understood. New technology facilitates improving workflow and better meeting the demands of customers. All of this requires regular change.

This regular change, however, must be planned, structured, and sold to the employees. It requires a culture where change is the norm. Where employees are engaged and involved in change. It requires leadership who support and encourage change, who are champions for change, and can sell their vision for change. Throughout this book a number of best practices are shared for managing change. Those best practices are the foundation for preparing for and benefiting from continuous change within the organization.

When engaging an organization's leadership in discussions around the need for planned, structured, and regular change, ask the questions shown in Table 9.1.

This sampling of questions, when asked of leadership, gets them thinking about what must be done in their organization to improve, to move ahead of the competition, and to better support customer needs—to, effectively, survive over the long term. The goal of these conversations is to get leadership to understand that through continuous planned change initiatives, they can stay ahead of the competition, attract and retain top talent, better meet customer needs, and improve in how the work gets done overall.

Table 9.1 A sampling of questions to engage leaders in understanding the need for continuous change

Industry-driven needs	
• What signs are you seeing in your industry that indicates the company may have to change?	• Which organizations are the best known in the industry and what are they known for among the competition?
Process-improvement needs	
• When were business processes last updated or refined in the organization?	• How have your current processes and procedures hindered your ability to meet organizational goals, solve problems, or meet customer needs?
Competition	
• Consider the competition. Who has surpassed your organization? In what way have they surpassed your organization?	• If a new competitor entered the market today, what advantage would they have over your organization?
Customer needs	
• What are the organization's biggest challenges with customers?	• What demands from customers are becoming more common?
Technology	
• Where can technology be implemented to better support the business?	• Where in the organization do pockets exist where technology is effectively and efficiently in use?
Employees	
• What can be put in place to enable the organization to attract and retain *the best* talent?	• Where in the organization are employees already implementing change regularly?

Organizations that have a culture of continuous change, evaluate that change regularly. If a change initiative that has been launched is not going well and is unlikely to be successful, it is terminated. This enables resources and budget monies to be routed to another change initiative that will be successful.

Organizations that have a culture of continuous change ensure that the following information, shown in Table 9.2, is true.

TYING CHANGE TO CONTINUOUS IMPROVEMENT

First, let's look at examples of rapid, urgent change and continuous improvement. Understanding the difference between the two enables development of a strategy around continuous improvement efforts and more effectively selling to the employees the need for continuous improvement.

Table 9.2   What must be true for a culture of continuous change

Is the following true of the organization?		
	• Leaders and managers understand that their roles include engaging their employees in change and helping their employees to adapt to change	
	• Leaders and managers understand that change is not easy for anyone and resistance is common	
	• Leaders and managers have a mindset of change and the skills and knowledge needed to lead and manage through change	
	• Employees throughout the organization—at all levels—understand the value of change and are encouraged to participate in change through suggesting areas of improvement	
	• Change is aligned to strategic goals, and those goals and that alignment is shared and understood by employees	
	• Regular roles and job descriptions supporting change exist in the organization —change team members, Stakeholder Support Committee members, change managers, etc.	
	• Performance management systems are aligned to a goal of continuous change and improvement	

(a) *A retail organization has lost more than 25% of their employ-ees to a competitor that has just entered the market. In fact, over the last year and a half, turnover rate has increased in a number of key functional areas. This problem is creating an impact on the customer side and the company's ability to get products out to market. In evaluating the situation, it was apparent that there were significant problems in hiring practices, performance management, technology, benefits, and compensation practices. All of these issues were contributing to the increase in employee turnover. The retail organization was sorely behind the competition in how employees were perceived to be treated. Leadership realized changes needed to happen and they needed to happen quickly! This required an urgent change. Key talent leaving for the competition would negatively impact the organization's bottom line in a very short period of time.*

(b) *In a management consulting firm, problem solving seemed to be a bit slower than usual. It took more time than usual to come to a resolution for internal problems. Processes used in the past didn't seem to be as effective as the organization grew and more people were involved. For some teams, it was unclear who the decision makers were. Leadership directed management to support their staff in taking the lead on evaluating and refining current processes for decision making in order to improve problem solving. As part of an overall continuous improvement initiative, leadership tasked all management with developing strategies for regular evaluation and refinement of processes.*

In Example (a), significant change must occur throughout the organization that would impact a number of assumptions under which the organization has been operating for a while. In this example, the organization has resisted efforts to effectively keep up with what is needed to retain top talent. The competition is perceived to take better care of employees and show their appreciation for efforts through better benefit and compensation programs. Failure to react immediately and with urgency will only serve to further impact the ability to source and retain talent, which will ultimately impact the bottom line.

In Example (b), there was no negative customer impact (yet) due to poor problem solving and decision-making processes. Rather, the organization is beginning to notice that processes are not as effective as they once were. Urgent and rapid change is not necessary (though it might be if the organization continued to ignore the situation and there was a bottom line impact). Rather, there is time to think through a solution and implement new or refined processes that will improve the situation. A smart organization will utilize this effort to kick off an evaluation of all processes toward a goal of continuous improvement in how the work is accomplished.

While the need for urgent change may not necessarily disappear completely in any organization, urgent change can certainly be reduced if an organization pays attention to what's going on internal and external to the organization and are *proactive* rather than *reactive*. Continuous improvement enables proactive change. Organizations that tend to see change as opportunities, that engage employees in regular change, and that embrace change are the organizations that can successfully implement continuous improvement initiatives. Through enabling and supporting small incremental changes, the organization regularly and continuously changes and adapts to the market, industry, and its customers.

Figure 9.1 provides a process for continuous improvement. Let's walk through Figure 9.1. First, the continuous improvement team identifies areas where improvement is possible. Identification might happen in any number of ways: through employee engagement survey data, customer feedback, observation, other feedback channels, or management direction.

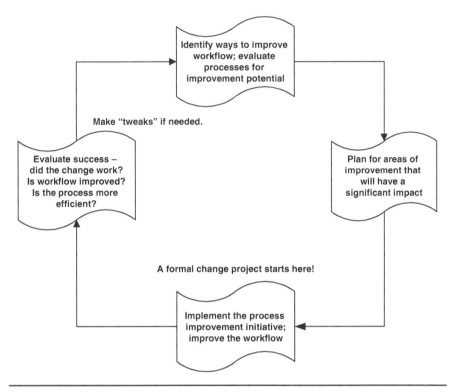

Figure 9.1   Continuous improvement cycle

For one of my clients, of the five change initiatives launched each year, three of them were continuous improvement initiatives. Areas for improvement for these three initiatives would be identified through suggestion boxes placed throughout the organization, employee engagement surveys, and through focus group meetings. This client launches five to eight change initiatives each year. Some of those initiatives are immediately launched, others are "reserved" for process improvement.

Once potential areas for improvement are identified, further study is done to determine if improvement in that area will have a positive impact on the organization. A positive impact may be reflected in a number of ways, including:

- Bottom line/return on investment
- Reduction in workload
- Shorter time-to-market
- Improvements in customer service

Once the client team had identified areas for improvement, they then dug deeper to determine areas of specific focus in order to have the greatest impact on both the organization and the individuals who were involved in working the process. This entailed getting those who were the most familiar with the process involved in understanding potential improvements that would be of value to everyone concerned and make the most sense, given the budgets available, the strategic focus of the organization, and the amount of time necessary to implement. Since the budget was allocated for three improvement initiatives, the team narrowed down the six processes identified for improvement to the top three, most impactful ones.

Once areas for improvement have been identified and accepted, this is the start of a formal change initiative.

The three process improvement change initiatives were officially launched. The members of the team who had been tasked with evaluating options for improvement were tasked with leading the three change initiatives. Additional employees were assigned to the project team, as well as to the change team (all of whom were familiar with the processes that were part of the change initiatives). Those involved in the continuous improvement initiatives were provided time from their "day jobs" in order to be involved in the initiative. This was considered to be a great opportunity for employees as it generated professional development and potential future career growth.

Once the improved process is initially rolled out to a smaller group and tested, it is evaluated to be sure it accomplished what it was supposed to accomplish. If successful, it is rolled out more widely within the organization. If necessary, it might be *tweaked* before it is rolled out to correct any issues. The cycle continues. Additional processes are evaluated as part of the organization's continuous improvement initiative (change regularly!).

Each year, the client changes the team involved in identifying improvement initiatives. The goal is to enable all employees to get involved in such initiatives. These are considered plum assignments and employees who are selected are those who are nominated to serve by their peers and managers and who have championed change in the past.

Not every organization thinks of continuous improvement initiatives as change initiatives. But they are! For organizations that regularly evaluate processes—even if only a small adjustment is made in a workflow—those are, effectively, change initiatives. Continuous improvement initiatives are beneficial in the organization by:

- Supporting and encouraging regular problem solving by employees
- Making gradual changes in how the work gets done
- Involving employees at all levels in the organization through using trial and error to find better ways of getting the work done
- Encouraging a mindset where employees regularly question how the organization works
- Improving communication and collaboration across the organization
- Reinforcing a focus on a collaborative and positive work environment
- Creating a proactive approach to evaluate and improve how work gets done

All of this encourages and supports continuous organizational change. And, it serves to engage employees in the long-term success of the organization.

Let's look at a brief example adapted from an initiative launched by a small company that provides at-leisure clothing to retail stores as part of continuous improvement efforts. Consider Figure 9.2.

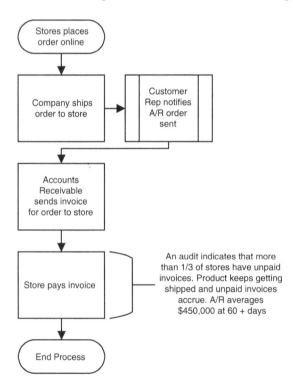

Figure 9.2   Process for order fulfillment—broken

As can be seen in Figure 9.2, the order fulfillment process is broken. There is no point where the customer representative checks to see if the store has paid previous invoices for orders. An audit conducted as part of strategic planning efforts showed that on average, the company carries accounts receivables for over 60 days at an average of $450,000. This has had a significant financial impact on the organization. In looking at change initiatives to launch, the senior leadership team decided that the process of order fulfillment needed to change in order to reduce the amount of outstanding accounts receivable. Figure 9.3 shows the process after a small change was made as part of a continuous improvement initiative.

In Figure 9.3, the process has changed to include a step where the customer representative verifies that the account is up-to-date with payments. If it is up-to-date, then the order is placed and shipped; if it is not up-to-date, the store is notified to pay past due invoices. If the invoices are paid in full, then the order is placed and shipped. If the store refuses to pay, then the account is turned over to Accounts Receivable, who would take care of the situation.

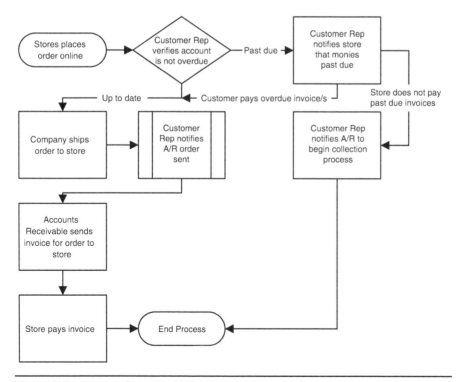

Figure 9.3   Process for order fulfillment—fixed

Given the success of this small improvement initiative, the at-leisure clothing company evaluates processes on a continuous basis to look for improvements. While this particular improvement had a significant, positive, bottom-line impact within three months of launch; the company will also launch change initiatives that have a far less impact. The company has found that even regular, smaller, positive impacts to the bottom line will have a long-term benefit.

Change and the Organization's Strategic Planning Sessions

If the organization is to move to a culture where change is the norm and employees strive for continuous improvement, change *must* be a part of strategic planning sessions. Consider this example:

> *A pharmaceutical company holds strategic planning sessions twice a year. One session is reserved for senior leadership team members only. The other session, however, is an all-day session that involves individual contributors from across the organization and some management-level staff. The first session includes senior leadership only and is focused on development of the strategic plan for the next few years. The second session, with select individuals and managers from across the organization, determines what change initiatives must be launched in order to ensure the success of the strategic plan. For example, leadership wants to move from research to development of a product. While this objective is a change in and of itself, there are a number of smaller changes that must happen first. These include:*
>
> - *Understanding the Phase 3 clinical trial process and the Food and Drug Administration (FDA) approval process*
> - *Understanding FDA requirements for approval of a product prior to development*
> - *Understanding resource needs to ensure a successful move from research to development*
> - *Understanding technology needs to support development*

In this example, in order to avoid such a large significant change becoming chaotic, this process enables validating a desired objective and determining the best way to move forward with the strategic initiative. Certainly, if it is determined that the organization is positioned to move forward with a strategic change initiative of moving from research to drug development; there are a number of smaller change initiatives that are a component of this much larger change. These include, but are not limited to:

- Creating an organizational structure to support drug development
- Creating processes behind submission of FDA documentation and applications
- Determining resource needs to support drug development

Organizational change initiatives must be aligned to the organization's strategy; therefore change will often come out of strategic plans. Successful organizations do not just change for the sake of changing; rather, they change to support the long-term strategy of the organization. When change is aligned to the strategic plan, it is more likely to have clarity and a vision that employees can understand and support. If change is *not* aligned to strategy, it is akin to throwing a dart blindly to see if it sticks.

Consider an organization that has a strategic plan that includes an objective of implementing a project management office (PMO). This same organization struggles with departments that are siloed; exhibiting a lack of collaboration and knowledge-sharing between departments. Launching a project to implement a PMO is a change initiative that cannot succeed *unless and until* other changes happen within the organization—namely, eliminating the silos between departments and increasing collaboration and knowledge-sharing which requires developing a team mindset in the organization.

ROTATING THE ROLE OF CHANGE TEAM MEMBERS

Change teams should be involved and contribute to the organization beyond working on a particular change initiative. The use of change teams for specific change-focused projects was discussed earlier in this book. Here, the focus will be on using change teams on a regular basis to engage in continuous improvement initiatives. Obviously, the larger an organization is, the more resources that will be available for change teams. As a best practice, have long-term change teams (as will be discussed in this section) work on continuous improvement initiatives; and specific project-focused change teams that work alongside project teams to implement larger change initiatives.

Consider the previous story about the pharmaceutical company—those individual contributors and managers who were invited to attend and participate in a strategic planning session were, effectively, change team members who were helping senior leadership to frame changes necessary to achieve a strategic objective. Figure 9.4 provides a path for selecting change team members to work on continuous improvement initiatives.

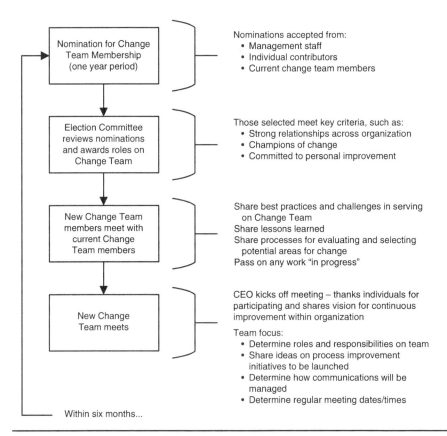

Figure 9.4   One potential path to select and kick off change teams

What Figure 9.4 does not reflect is that the change team—at three month, six month, and nine month intervals—meets to share what is working and what is not (lessons learned captured) as well as evaluate the effectiveness of their continuous improvement efforts. Within six months of the change team kicking off their work, the process starts over again with nominations requested for change team membership. Figure 9.5 provides an example of a nomination form that might be used to nominate change team members.

Figure 9.6 provides an example of an evaluation form to select change team members.

For a number of organizations, interviews are a major component of selecting change team members. In evaluating nominations, it is important to keep in mind that a diverse group is needed to serve on the change team. This includes:

Nomination Form: Change Team Membership		
Candidate Name: \<name\>	**Candidate Department Affiliation:** \<department name\>	**Nominator Name:** \<name\>
Instructions: Be specific in responding to each question below; providing examples to support your response.		
Why should this candidate be selected to serve on the Change Team?		
What can this candidate bring to the Change Team that would be unique?		
How has this individual embraced change?		
How has this individual helped to champion or support change within the workgroup, department, or organization?		
How has the individual demonstrated the ability and willingness to lead change?		
When has this individual demonstrated the ability to serve in both a team leader and a team member role?		
Which of the following skills and competencies has the individual displayed?	☐ Team leadership ☐ Strong communicator ☐ Presentation skills ☐ Strong problem solver	☐ Conflict management ☐ Negotiation ☐ Strong collaborator ☐ Team player ☐ Other: _____
What else do you want to share about this individual that would support their being nominated to serve as a member of the Change Team for the upcoming year?		

Figure 9.5 Nomination form: change team members

- A variety of experience levels and roles represented
- Representation from across the organization
- A variety of cultural backgrounds
- A variety of personalities

Rotating change team members provides opportunities for employees to serve throughout the organization. Of course, a smaller organization may not have enough people to rotate membership regularly. A larger organization, however, should consider rotating membership on at least an annual

Evaluation Form: Change Team Membership		
Candidate Name: <name>	**Candidate Department Affiliation:** <department name>	**Evaluator Name:** <name>
Should this candidate be selected to serve on the Change Team? Be specific beyond replying with "yes" or "no." (Consider nomination forms received, interview with candidate, personal experiences, skills and competencies of individual, etc.)		
If you responded "no," what specifically might this individual do to be considered for a role on the Change Team at a future date? (If none of the options on the right apply, please be specific in the Comments section below.)	☐ Gain more experience in organization (too junior) ☐ Increase collaboration across organization ☐ Show willingness to personally change ☐ Improve communication skills ☐ Gain more experience working on cross-functional teams ☐ Gain more experience working on change initiatives within workgroup/department ☐ Other: _____	
Comments:		

Figure 9.6   Evaluation form: selecting change team members

or biannual basis. Of course the period that change team members serve should be linked to the type of continuous improvement efforts that they are expected to undertake. Smaller efforts may permit rotating change team members on a quarterly or six-month basis. Larger efforts may mean that individuals serve for one or two years in the role.

> *I know of one smaller organization that has only 100 employees. However, this organization still launches change teams! There is no formal process to be nominated and selected; rather, individual employees volunteer to serve on the change team. Change teams serve for 18 months. They focus on continuous improvement in the following areas: marketing, human resources, finance and administration, product development, customer relations, and external partnerships/vendors.*

ENDING ONE INITIATIVE AND STARTING THE NEXT

In order for continuous improvement to become the norm in the organization, there needs to be a clear roadmap, as well as detailed processes and procedures for identifying, planning, implementing, and evaluating continuous improvement initiatives. It cannot be haphazardly accomplished. An organization does not want teams identifying, implementing, and evaluating processes differently year after year. There needs to be consistency in how process improvement is done in the organization. Remember that continuous process improvement is cyclical, must be measurable, and should be incremental. An organization does not want to initiate one project that looks to improve *every* process in an organization. Pick one process, improve it, evaluate if it was successfully improved upon, and then start again (remember Figure 9.1). Continuous process improvement enables a regular, proactive, and disciplined approach to collect information about how work gets done, evaluates the way the work is getting done, and makes incremental improvements and tests those improvements to ensure they are positive. Figure 9.7 is one potential roadmap for continuous improvement efforts in an organization.

Table 9.3 provides a list of potential processes and procedures which should be considered for consistency in moving to a culture of continuous improvement.

An organization just starting such an initiative might choose to assign the first team with the project of determining processes and

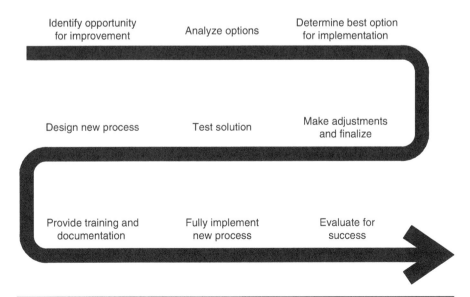

Figure 9.7   A roadmap for continuous improvement

Table 9.3 Processes and procedures for continuous improvement consistency

• Rotating team selection process	• Use of benchmarks and industry data
• Status reporting and team meetings	• Use of technology to support collaboration
• Process improvement audit procedures	• Capture and share lessons learned
• Procedures to evaluate current processes	• Procedures for testing and selecting solutions
• Processes for identifying areas of improvement	• Processes for development of training and documentation
• Procedures to evaluate impacted areas: resources, technology, etc.	• Processes and channels for gathering and evaluating feedback
• Communication channels to be used	• Processes for making decisions and resolving conflicts

procedures to be used for all future process improvement initiatives. This team should be responsible for determining and setting guidelines around:

- *How to determine the impact of a process improvement change*
- *How to validate the impact of the change*
- *How to communicate about the change and its value*
- *How to test the change on a smaller scale to reduce risk*
- *How to measure and evaluate success of the change once implemented*

Continuous improvement is an approach to testing changes to see how effective and valuable they are. In order to reduce any potential negative effects from improvement initiatives, a change may be made and tested on a smaller scale—possibly with one workgroup or team—and then, if successful, rolled out to a broader group to implement. This enables making changes relatively risk-free in the organization.

As with any change initiative, communication in continuous improvement initiative undertakings is absolutely essential. For consistency, regular channels for communicating should be implemented and used with every group working on continuous improvement. There must be systems in place to track initiatives, share best practices and knowledge, and capture and share lessons learned.

The benefits in having a culture of continuous improvement is not just for the organization—employees benefit too! Continuous improvement establishes a better work environment overall.

Smaller organizations may find value in getting the *entire* organization involved in process improvement initiatives. With smaller numbers of employees, it is much easier to engage everyone in the process. Ask for ideas on how to improve workflow based on:

- Employee's past work experiences
- Employee's readings or research done
- Employee's networks

It is easier for an organization to *start* with a culture of continuous improvement early on in the life of the company and then continue that culture as the organization grows. However, for larger organizations, it is certainly possible to get started with a culture of continuous improvement. Consider all the best practices shared in this book so far on implementing change and follow those to a successful culture of continuous improvement.

Let's end this chapter with another client example. Consider how one global manufacturer started an initiative of continuous improvement and, on an ongoing basis, effectively manages continuous improvement efforts in the organization, realizing regular improvement in how the work gets done:

A Chief Manufacturing Officer in a global manufacturing company started an initiative about six years ago to engage employees in finding ways to get work done more efficiently and effectively. The initiative challenged employees to look at an area of the business where improvements could be made and to develop a proposal for leading that improvement project. The winner of the initiative would be relieved from their daily responsibilities in order to lead the initiative, as well as be provided additional resources and funding to implement the improvement project. By the due date, a total of 85 proposals were received from employees at all levels throughout the organization and representing all geographic locations. While the Chief Manufacturing Officer had hoped he would have interest, he never expected this level of interest! Three winners were selected. However, the other proposals also had merit. The Chief Manufacturing Officer, after a conversation with his peers on the executive team, made the following announcement: "Given the tremendous number of ideas we received to improve how work gets done, the leadership team will be launching a strategic project to develop a roadmap for the implementation of continuous improvement initiatives in the organization."

This was the start of the organization's move toward a culture of continuous improvement. While the employees were thrilled with the idea of identifying and implementing process improvement initiatives, many line managers were less than excited. The project had two teams: the first was a project team to manage the day-to-day project tasks—the second was a change team that was comprised of managers. These managers who comprised the change team were thrilled with the prospect of employees taking the initiative in process improvement. Their main task was to engage all other managers who were less than thrilled with the initiative.

The reader will recall that change teams should be comprised of a variety of individuals—rarely only management. However, in this case, managers were essential on this team because it was *management* who was resisting the change. The best way to influence them was to have their peers—who were supportive of the change—share the value and benefits of the initiative for management. *Before* the project team began their work, the change team of managers who championed the initiative was deployed across the organization to convert their peers from resisters to supporters. This was a four-month effort. Then, the project team began their work.

At the end of the initiative to create a culture of continuous improvement, deliverables included:

- *A roadmap for continuous improvement initiatives*
- *Processes and procedures for identifying and selecting areas for improvement*
- *Processes and procedures for nominating and selecting individuals to serve on the team*
- *Detailed roles and responsibilities for individuals who served on the process improvement team as well as terms of service (teams would consist of 8–10 individuals who would serve two year terms; each year, 4–5 team members would be rotated on and off the team so that there was consistency for longer term process improvement efforts, which meant that the first year the team started some members would only serve one year)*
- *The launch of a portal for collaboration, communication, and capturing information about process improvement initiatives*
- *Development of a variety of templates and documentations to support process improvement efforts*

Prior to starting process improvement initiatives, training was provided to all employees via a combination of virtual and face-to-face lessons on:

- *Effective communication across cultural boundaries*
- *Decision making*
- *Resolving conflicts*
- *Virtual teaming*

In order to keep managers engaged and supportive of improvement efforts launched by their employees, a change was made to compensation and bonus packages for managers that promoted continuous effort by ensuring management:

- Enabled employees to work on process improvement initiatives (by relieving them from day-to-day responsibilities)
- Encouraged and championed process improvement in their specific areas of responsibility
- Improved collaboration across divisions and departments through assisting employees in identifying and finding cross-functional opportunities for improvement

The goal was not that managers would lead such efforts, but rather take a back seat and enable their employees to determine where improvements are needed and to lead the effort in implementing the improvements.

This is just one example of how one organization enabled a culture of continuous improvement. This did not happen overnight; rather it took time to get to the point where improvement became a regular part of the organization. As with any major change, time, resources, and money must be invested to achieve success. While the organization had a number of challenges, they were resolved quickly. Structures put in place, including changes to compensation and bonus packages for both managers and employees, caused increased commitment to seeing the leaderships' vision come to realization.

This book has free material available for download from the
Web Added Value™ resource center at *www.jrosspub.com*

10

GETTING STARTED

"Change almost never fails because it's too early.
It almost always fails because it's too late."
Seth Godin

In today's global competitive marketplace, those organizations that adopt a culture of continuous change will survive and thrive. To compete in a crowded marketplace, organizations must learn, adapt, and innovate *continuously*. This requires creating a culture of continuous change. To do this, however, the employees of the organization must be comfortable with risk, making decisions rapidly, and changing how work is done—the status quo should be something to be avoided.

Any organization is capable of being one that looks at change as a regular, proactive part of the organization. I have seen the smallest organizations regularly evaluate how work gets done to ensure continuous improvement. One of my clients—a global organization of over 10,000 employees—launches six to eight process improvement initiatives each year that keep the organization immersed in continuous change.

However, some organizations are further ahead in accomplishing this objective than others. Developing a culture of regular, positive, and proactive change requires:

- Calculated risk taking (mistakes are going to happen)
- Developing processes and procedures for consistency in the implementation of change initiatives
- Utilizing systems to identify areas of improvement, as well as to plan for, implement, test, and measure the success of change initiatives

- The development of criteria as well as a selection process for ensuring the right initiatives at the right time, that will have the most positive impact
- Developing metrics and measures to evaluate the success of changes made
- Developing feedback and evaluation systems
- Ensuring a variety of communication channels

The most successful organizations realize that they must continually change. Without employee support, continuous and proactive change is just not possible. Leadership alone cannot initiate changes and certainly in a culture of continuous improvement, much of those areas identified for improvement come from the employees—those individuals who are working with the process day after day. Moving toward positive and continuous change requires understanding challenges within the organization—those that exist at the executive level and those that exist for senior management, middle management, supervisors, and individual contributors. This requires frequent and regular communications up, down, and across the organization as well as engaging stakeholders—everyone in the organization—in conversations around change.

The goal of adopting a culture of positive and continuous change requires commitment from the top; and it requires a strategy. The executives in an organization can't simply decide they are going to have a culture of continuous change. A strategy must be developed and executives must sponsor and support such a strategy. The executives are the individuals who will commit resources from their own staff, as well as allocate budgets. The next level down of leadership must also commit; as must mid-level managers and supervisors. These are the individuals who will need to champion change. They will be the ones who will encourage and support their staff to embrace change and be proactive in making strides toward continuous improvement. They will provide training to enable employees to be successful. And, of course, employees must believe that leadership at all levels will support them, accept that mistakes will happen, and that errors in judgement might be made. While processes can be put in place to reduce the impact of risks to the organization, mistakes will still happen. Employees must know that they won't be penalized for trying something new.

> *Certainly organizations can and should reduce the impact of mistakes by having clear processes in place for resolving issues and making decisions. By ensuring regular feedback channels as well as*

capturing and applying lessons learned, mistakes can be avoided or reduced. Additionally, ensuring that larger change initiatives are broken down into smaller projects enables evaluating and testing as changes are made, thereby reducing risks overall. Even when employees lead continuous improvement efforts, leadership should be regularly apprised of the status of the initiatives.

In this chapter, the focus will be on developing the strategic plan for making an organization one in which change is a norm—where employees from throughout the organization, at all levels, contribute to the growth of the organization through identifying, planning for, and implementing regular, positive change.

DEVELOP THE STRATEGIC PLAN FOR MAKING ORGANIZATIONAL CHANGE A REGULAR PART OF A GROWING ORGANIZATION

There are any number of books available on strategic planning. This chapter does not purport to be a primer on it. Rather, it provides a number of steps to get started in developing and implementing a strategic plan for an objective around continuous change in the organization.

A focus on continuous and positive change requires an understanding of the current culture and its strengths and weaknesses. If any weaknesses are going to impact the ability to make change a regular occurrence in the organization, those weaknesses must be addressed *before* the organization can launch a strategy to move to a culture where change is embraced.

A colleague of mine shared that her organization started with a number of strengths that enabled moving to a culture of change. These included: employees who were engaged and committed to the success of the organization, knowledge that was regularly shared, and employees who could share the vision for the organization as if they devised it themselves. Additionally, management regularly reached out to employees to ask for their ideas and suggestions on how to improve the organization.

On the flip side, an organization where employees are not engaged or do not understand the vision, and where employee turnover is common, would find it very difficult to implement a strategy to make organizational change a norm. In these organizations, leaders rarely, if ever, reach out to employees for ideas and suggestions. Such organizations tend to be hierarchical with decision making left to the highest levels in the organization. Much needs to change long before the organization can consider taking a strategic viewpoint of continuous change if they are to be successful in their efforts. Consider the questions posed in Figure 10.1.

This figure provides a number of key questions to consider regarding the readiness of the organization to implement a strategy that will enable change to be the norm. These questions are similar to ones asked to determine the readiness for change overall (shared earlier in this book). Certainly a handful of *no* responses does not mean an organization should shy away from working toward change as the norm, but rather indicates areas that *must first be improved upon prior* to implementing an overall strategy of regular and consistent change.

In responding to the questions posed in Figure 10.1, a medical devices company realized that until they could get employees to proactively look for better ways to get work done, they could never move to a culture where change is the norm. In researching the primary reasons for a lack of motivation in proactively looking to improve, leadership noted that risk taking was not encouraged and, in fact, those employees who had taken risks were punished when something went wrong. Additionally, decisions were rarely left in the hands of employees. It seemed that even the most basic solution to a problem had to be voiced by leadership. The project leadership wanted

Can "yes" be answered for each question?	
	Are employees engaged in the organization?
	Do employees frequently look for better ways to get work done?
	Is collaboration common in the organization?
	Do employees share knowledge and best practices cross-functionally?
	Does leadership support and champion change?
	Are retention rates high?
	Are employees held accountable?
	Are employees enabled to take calculated risks that enable for growth?
	Do employees show initiative?
	Do employees embrace the vision and mission of the organization?
	Are decisions made based on the organization's vision and mission?
	Are decisions able to be made at the lowest levels of the organization?

Figure 10.1 Is the organization ready for change as the norm?

to launch was to design and implement a system that rewarded individuals for taking risks in improving how work gets done, as well as enabling decision making at lower levels. This would require changes to compensation and reward systems, as well as training employees and managers in risk-taking best practices and decision making. It would also require processes and procedures around making decisions. However, prior to even implementing that project, it was necessary to dig deeper into why risk taking was discouraged and why leaders felt they, and they alone, could make decisions. Then, it was necessary to increase the comfort level of managers in, effectively, letting go.

Strategic planning is about setting priorities, making decisions, and taking actions to meet objectives—then measuring the success of the actions taken. Strategic planning answers three questions:

1. Where is the organization today? (*Present*)
2. Where does the organization want to go? (*Future*)
3. How does the organization get there? (*Present → Future*)

Figure 10.2 provides a four-step process for strategic planning. This is just one potential process.

As is apparent in Figure 10.2, there is much that is done up front *before* actually producing the plan. Of particular importance—and where a decision is made to move forward or not with the strategic plan objectives—is Step 2 and Step 3. Step 2 is where challenges that may exist that may impact the objectives are explored further. Challenges are not necessarily an indicator that the strategic initiative to be an organization focused on change should be halted, but rather, exploring those challenges will enable for determining changes that must happen so that the organization can more easily adopt a future culture of regular and continuous change. Let's walk through this four-step process with an example.

A regional health and wellness organization wants to start implementing continuous improvement initiatives. From the perspective of the Chief Executive Officer (CEO), this is a perfect time to move forward with this strategic initiative. The organization, which started with one location and 10 employees three years ago, now has three offices with 85 employees. Table 10.1 provides a brief overview of what they accomplished in each step in developing a strategic plan for continuous improvement and engaging employees in the process.

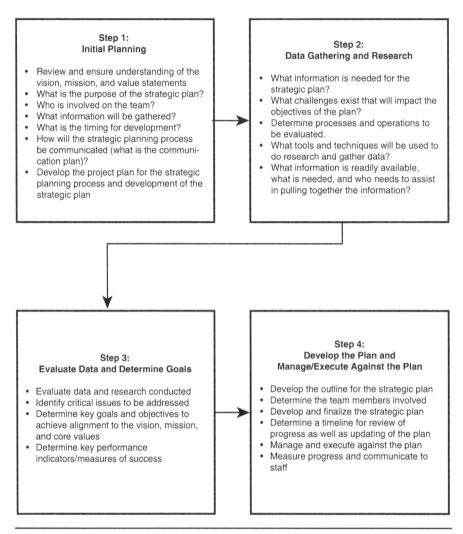

Step 1:
Initial Planning

- Review and ensure understanding of the vision, mission, and value statements
- What is the purpose of the strategic plan?
- Who is involved on the team?
- What information will be gathered?
- What is the timing for development?
- How will the strategic planning process be communicated (what is the communication plan)?
- Develop the project plan for the strategic planning process and development of the strategic plan

Step 2:
Data Gathering and Research

- What information is needed for the strategic plan?
- What challenges exist that will impact the objectives of the plan?
- Determine processes and operations to be evaluated.
- What tools and techniques will be used to do research and gather data?
- What information is readily available, what is needed, and who needs to assist in pulling together the information?

Step 3:
Evaluate Data and Determine Goals

- Evaluate data and research conducted
- Identify critical issues to be addressed
- Determine key goals and objectives to achieve alignment to the vision, mission, and core values
- Determine key performance indicators/measures of success

Step 4:
Develop the Plan and Manage/Execute Against the Plan

- Develop the outline for the strategic plan
- Determine the team members involved
- Develop and finalize the strategic plan
- Determine a timeline for review of progress as well as updating of the plan
- Manage and execute against the plan
- Measure progress and communicate to staff

Figure 10.2   A four-step process for strategic planning

While the CEO was anxious to move forward with outlining a strategy to launch process improvement initiatives on a continuous basis, there was significant concern among leadership overall that several processes were undocumented, poorly documented, or worse yet, there was limited understanding of how employees got from A to Z. Therefore, the initial strategic plan outlined an initiative to document all current processes "as is." Once this initial project was completed (which was expected to take six to nine months in

Table 10.1   Organization example: what was accomplished at each step

Step	Example—Moving Through the Four Steps
1	• Develop and share the vision for continuous improvement and align to other long-term strategic goals • Select employees who will assist in steps 2 and 3 (Research Team) • Develop a list of information to be gathered—prioritize business processes for evaluation • Develop early communication plan to share upcoming desired strategic initiative throughout organization
2	Conduct research: • Survey employees to determine areas of focus for continuous improvement (employee perspective) • Review past employee engagement surveys • Review best practices within and external to industry • Evaluate processes which have not been updated within last few years Challenges expected: • Most all processes were poorly documented and others undocumented • A few of the individuals with expertise in particular processes had left the organization
3	• Many key issues identified in current documented processes; many processes inconsistently applied • Some key processes undocumented; insufficient information among employees to effectively document the entire process (employees know "bits and pieces" of the process) • Key performance indicators (KPIs) not in use within organization; limited understanding among leadership as to the value of utilizing KPIs • Initial KPIs proposed (for further discussion) included: Key Customer Satisfaction, Customer Loyalty Index, Number of Customer Referrals, Employee Satisfaction, Dollars Saved by Employee Suggestion, Employees on Self-Managing Teams, Time Spent on Quality Improvement Activities, Improvement in Productivity
4	• Outline for strategic plan developed • Finalize list of prioritized process improvement initiatives (factors to prioritize: return on investment (ROI), customer impact, improve workflow, improve collaboration) • Confirm first project: documentation of *all* current processes "as is" • Communication plan developed to share outline of strategic plan throughout organization • "Road shows" scheduled (visits to each office to discuss initiative) • Initial Process Improvement Teams created (representative of all functions and all offices) • Processes and procedures outlined for working on continuous improvement initiatives (for review and further refinement by initial Process Improvement Teams)

duration), then continuous process improvement initiatives would begin. In preparation for this launch, while processes were being documented, work would begin on determining factors for prioritizing and selecting process improvement initiatives, development of processes and procedures for work on such initiatives and creation of the first process improvement teams to work on the initiatives.

Strategic planning is a process that includes inputs, activities, and outputs as shown in Figure 10.3.

Inputs, activities, and outputs can be aligned to the four-step process for strategic planning (Figure 10.2). The first three steps are part of inputs and activities. Step 4 is the output.

Table 10.2 provides a sampling of questions to ask of employees to get their input for the development of the strategic plan.

Let's look at another example. One key tool that should be used is a strengths, weaknesses, opportunities, and threats (SWOT) analysis. This is vital for strategic planning. Conducting a SWOT analysis as part of strategic planning will make it possible to find the best path for moving forward to achieve goals, as well as visualizing those challenges (weaknesses) that may impact the ability of the organization to be successful. Figure 10.4 provides an example of a SWOT analysis for an organization that is trying to achieve success in process improvement initiatives.

First, some background. In this organization, one strategic objective launched two years ago was to increase the number of continuous improvement initiatives launched and implemented. The organization wanted employees to regularly change how work efforts were accomplished, utilizing new technologies, best practices, and refining processes in order to meet

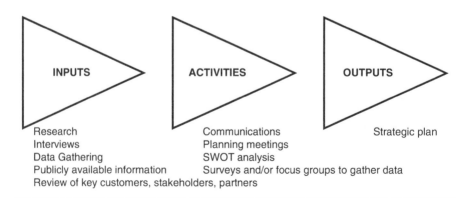

INPUTS

ACTIVITIES

OUTPUTS

Research
Interviews
Data Gathering
Publicly available information
Review of key customers, stakeholders, partners

Communications
Planning meetings
SWOT analysis
Surveys and/or focus groups to gather data

Strategic plan

Figure 10.3   Inputs-activities-outputs

Table 10.2   Sampling of questions to engage employees in developing the strategic plan

• Where do vulnerabilities exist in the organization that may jeopardize success? • What trends do you see in the marketplace that may have a negative impact on the organization? • Where do you believe the organization should focus its resources and investment? Why?	• What interests you about the work the organization does? • What is most frustrating to you about working in the organization? • Based on feedback or comments you hear from customers, where do you think the organization should focus its strategy? • Do you believe the organization lives by its vision, mission, and core values? Why or why not?

the needs of a customer base that was rather fickle. About a year ago the organization realized that, while senior leadership pushed for continuous improvement and engaged employees in the initiatives, five out of every eight initiatives launched were not successful. In fact, over the last six months a number of unsuccessful initiatives had a significantly negative impact on the customer base.

The information in this SWOT analysis (Figure 10.4) has been reduced to only show the components relevant to the desire to increase the success of improvement initiatives undertaken within the organization.

Based on the information in the SWOT analysis, senior leadership put the strategic objective of continuous improvement on hold and undertook the following initiatives:

- Investment in collaboration tools (new technology)
- Development of processes for implementing continuous improvement initiatives
- Development of project management best practices, including risk identification and management
- Revision of leader and manager performance and award systems to put a focus on supporting continuous improvement

Additionally, training was developed that would be utilized by new hires through to senior leadership to ensure success in continuous improvement initiatives within the organization. In this example, let's look further at details around developing the strategic plan. First, strategic plans developed in the organization must be aligned to the mission, vision, and core values of the organization. The mission is actually the beginning of the strategic

STRENGTHS	WEAKNESSES
• Willingness among employees to undertake continuous improvement initiatives • A quarter of senior employees have experience in project management • Senior leadership has supported continuous improvement • Lower level managers have supported employees' work on process improvement initiatives	• Technology is outdated (no collaboration tools) • There is no process to apply lessons learned • There is limited training in identifying and managing risks • Managers are sometimes too "hands off" • Mid-level leadership have not consistently supported continuous improvement • Limited project management best practices in place • Many process improvement initiatives launched were done with 0 budget • There is no consistency across the organization in launching and implementing initiatives
OPPORTUNITIES	THREATS
• Senior leadership interested in improving situation • Budget money is available to invest in technology • New employees hired in last three months have extensive experience in process improvement • A recently secured business loan has a percentage available to enable for improvements in how work is done • New mid-level leadership hired in past year expressed an interest in improving the situation	• Customer complaints have risen and new competition poses the threat of taking away customers • A major competitor has improved their time-to-market for new products and has increased innovation (a press release attributed this to improvements in how work is accomplished) • Weariness beginning to set in with employees regarding the ability to be successful in continuous improvement activities

Figure 10.4 SWOT analysis example

plan and the end is progress toward achieving the vision of the organization. The core values help in the development of strategies and tactics to achieve the objectives of the strategic plan. It may sound confusing, but hang in there—it will become clearer.

- A mission is a very brief description of the organization's purpose— why it exists.
- A vision is what the organization hopes to be in the future—what it strives for.
- Core values define how the organization's employees will engage with and support customers, employees, and stakeholders. They speak to the integrity and ethics of the organization and what is considered appropriate or inappropriate behavior.

Let's assume a health and fitness organization's mission is to *enable members to be strong and physically fit and live healthier, happier, and longer lives* and its vision is to *be the premier health and fitness organization in the nation, with a focus on the needs of each individual member.* Its core values include:

- Treat everyone with respect
- Only make promises that you can keep
- Communicate openly and honestly
- Provide superior customer service; meet the needs of each individual member

The senior leadership team wants to create a culture where change is the norm and employees in all locations regularly evaluate and initiate change to better meet the needs of members. This decision was made due to recent complaints that the clubs have not been updated, membership benefits have been stagnant, and a number of other similar complaints. When the initiative is launched, each location develops strategies and tactics to meet the objective of creating a culture where change is the norm. Table 10.3 provides a sampling of some strategies and tactics developed by the membership and catering departments in particular.

> *Strategies answer the question, "How will we accomplish the organizational objective?" and tactics provide specifics on how that strategy will be achieved.*

As can be seen in Table 10.3, the membership and catering departments have developed specific strategies and tactics to help them achieve a goal of creating a culture of continuous improvement. When implementing change—such as in this example—each division, department, and workgroup or team must determine what that change specifically means for them. What is important in ensuring success in implementing a strategic plan is that every department is aligned with the next. In Table 10.3, for example, both marketing and catering have a tactic of surveying membership. Imagine the scenario where both marketing and catering send out separate surveys to membership. Undoubtedly, this would impact members' perception of the organization. Rather, each department (in this case, marketing and catering) would share across the function their departmental strategic plans and tactics to achieve the organizational strategic objective of creating a culture of continuous improvement. Through sharing, it would be obvious that a survey was a tactic for these departments and would have therefore been coordinated so that *only one* survey would be

Table 10.3   Example strategies and tactics to achieve a strategic objective

Strategic Goal: Create a Culture of Continuous Change/Improvement		Department: Membership **Department Goal:** Improve member experience
Strategy		**Tactics**
Determine strengths and weaknesses at clubs		• Survey membership • Review employee engagement data from club employees • Conduct focus groups of members • Conduct focus groups of employees who work at clubs

Determine differentiators with competition		• Research
		   • Membership benefits at competitor clubs
		   • Review industry reports

Strategic Goal: Create a Culture of Continuous Change/Improvement		Department: Catering **Department Goal:** Develop healthier catering options to meet needs of diverse membership
Strategy		**Tactics**
Survey membership		• Survey membership on catering expectations/needs/wants
Research		• Review nutritional studies • Hire consultant to assist in revising menu offerings

sent to members including questions that meet the research needs of each department.

For this organization, the Human Resource (HR) department was responsible for engaging all employees in the change. Table 10.4 provides a couple of HR's strategies and tactics.

Similar to Table 10.3, the strategies and tactics in Table 10.4 are not a complete list of what might be undertaken in each department. In Table 10.4, HR is focused on communication to engage employees in change and in reviewing and researching how the organization must look (organizational structure) as well as what systems need to be refined (for example, performance management systems) in order to support a culture of change.

Key performance indicators (KPIs) should be established to evaluate the success of employees in achieving the organizational strategic goal—in this example, creating a culture of continuous change.

Table 10.4 Example of Human Resource's strategies and tactics

Strategic Goal: Create a Culture of Continuous Change/Improvement	Department: Human Resources **Department Goal:** Engage employees in change initiative
Strategy	**Tactics**
Communication	• Develop strategy for communicating on change using a variety of channels • Launch communication and collaboration portal • Schedule meetings to discuss strategic objective of culture of continuous improvement
Refine organizational structure and systems	• Evaluate how current culture supports/does not support a culture of continuous improvement • Research/gather data on organizational structures that support continuous change cultures • Draft restructure of organization based on best practices • Evaluate changes to performance management system to reward and compensate employees based on a culture of continuous change

KPIs are metrics that evaluate how effectively the organization is achieving its strategic goals. KPIs may include, for example, customer acquisition, number of customer referrals, or complaints not resolved on first call.

In the example of the health and fitness organization, KPIs to measure success in achieving a culture of continuous change might include:

• Key member satisfaction
• Member wait time for access to fitness equipment
• Number of ideas submitted by employees weekly
• Number of membership complaints
• Number of process improvement initiatives launched monthly

Let's review one more tool—Porter's Five Forces—created by Michael Porter of Harvard University. It is a framework that analyzes the organization's competition and, therefore, profitability and viability of the organization within its industry and the marketplace. It is another tool (activity) in an organization's toolbox to gather information for the strategic plan. Table 10.5 provides a brief review of each of the forces in Porter's Five Forces model.

Table 10.5 Brief description: Porter's Five Forces

Force	Brief Description
Competition	This force includes the number of competitors in the industry as well as how similar the products and services are (in quality, pricing, etc.). The more competition that has similar products and services as the organization, the more competitive the marketplace in trying to engage customers in buying products and services.
New Entrants	This force looks at the possibility of additional competition entering the industry. If it is easy to launch a business (limited regulations, limited investment needed, etc.) the more likely that competition will increase.
Suppliers' Power	This force looks at how easy it is for an organization to change suppliers. If there are limited suppliers, then it is not easy for an organization to work with another supplier and therefore the supplier has more power than the organization (which could impact the cost of supplies).
Buyers' Power	This force looks at the power that buyers (the organization's customers) have to impact pricing of products and services. If an organization has a few buyers that drive much of the revenue for the organization, that gives the buyers more power over the organization. If competition has similar products and services, then buyers have more options which also gives them more power over the organization.
Substitute Threat	This force looks at the number of substitutions that exist for an organization's products and services. If a buyer can purchase a substitute that will serve the same purpose, costs less, and is readily available, the power of the organization in the marketplace is reduced.

For more information about Porter's Five Forces, see *Competitive Strategy: Techniques for Analyzing Industries and Competitors* by Michael E. Porter.

These are just two tools that organizations might use in order to get an understanding of their organization and where it exists within their industry and marketplace. The information from these tools—as well as others utilized by organizations—helps to design a disciplined approach to determining how to move toward a culture of continuous change. These tools enable for identifying initiatives that must be launched *before* trying to move to a culture of continuous change.

Culture is the reason that many organizational change efforts fail. Given that, it is obvious that the current culture must be examined when implementing an initiative to move toward continuous change. If the current organizational culture does not support continuous change—for example, the organization values the status quo—the culture must change first. Culture change takes time and effort; it does not happen overnight. But it is possible! In order to accomplish this, it is not sufficient for the CEO or other senior leadership to demand change; they must influence others in the organization to come along with change. This is not done in an all-staff meeting but rather in smaller group meetings and forums where employees can share ideas and discuss challenges. In getting started toward continuous change within the organization, executives and other leadership should begin to have conversations around change and its value to the organization, as well as the individual employees who comprise the organization. Ideas for doing so have been shared earlier in this book. A template for a strategic plan is available for download from the Web Added Value™ Resource Center at www.jrosspub.com/wav.

Who Should Be Involved?

While certainly executive leadership must lead the strategic effort to move toward a culture of continuous change, many others from throughout the organization must be involved. As shared earlier in this book, senior leadership alone cannot drive or sustain change—and certainly they cannot drive implementation of a strategic plan on their own; they need employee participation. Employees from throughout the organization must be engaged if a change initiative—especially one as challenging as creating a culture of continuous change—is to be successful. Getting a broad group of stakeholders involved in the strategic planning allows for increased awareness of the process (including how the organization gets from Point A to Point B), understanding of the value, and commitment to help implement the objectives set.

Figure 10.5 provides an overview of roles involved and key responsibilities of those roles when moving toward a culture of continuous change.

As is seen in Figure 10.5, roles from throughout the organization are involved in a variety of ways to achieve a goal of change as the norm by staying engaged in the initiative and providing feedback—from the executive and senior leadership team who are responsible for developing and sharing the vision and providing resources and budgets, through to individual employees who are responsible for doing the day-to-day work.

When moving toward a culture of regular and proactive change, it is essential to get *everyone* in the organization involved in the effort, regardless of their status, role, or responsibilities.

> *One organization of over 2,600 employees launched a strategic initiative to engage employees in continuous change. The CEO deployed change agents from throughout the organization to engage every employee in discussions around the benefit of continuous*

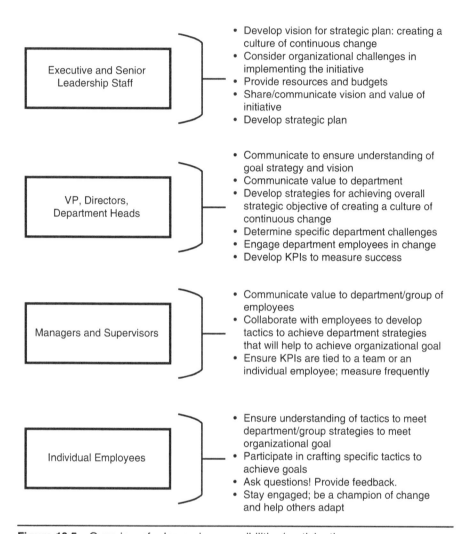

Figure 10.5   Overview of roles and responsibilities/participation

change. Since the organization had multiple shifts, change agents met with employees during various times throughout all shifts that were convenient for employees. This resulted in increased participation in shaping the initiative and made employees feel valued. Additionally, by engaging these change agents and having conversations with employees, a number of challenges that the CEO and other leadership were not fully aware of were brought to light. These challenges were addressed prior to the strategic initiative being

*launched. Without knowledge of these challenges, the strategic ini-
tiative would have been launched and would likely not have been
successful. By addressing these challenges, employees felt more engaged in the initiative, more comfortable with the strategic goal of
the organization, and felt that their opinions and ideas mattered.*

A stakeholder management matrix helps develop an understanding of
each stakeholder group that will be impacted by the initiative. Certainly,
an initiative that involves creating a culture of continuous change impacts
every stakeholder in the organization. Figure 10.6 provides an incomplete
example of a stakeholder management matrix.

Stakeholder Management Matrix					
Stakeholder Group	Area of Concern	Needed/Desired Level of Involvement in Initiative	Expectations of Change Leadership	Initial Communication/ Engagement Recommendations	Responsible Change Team Member(s)
TIMEFRAME: *Prior* to Launch of Organizational Initiative					
EXAMPLE: Human Resources	*Impact to performance management system* *Silos between departments*	*Significant involvement* *Need assistance in engaging employees in initiative throughout implementation* *Need data around specific impact to performance management system as well as design for new system*	*Need better understanding of the "why" of this initiative* *Need support in leading this change from an HR perspective (limited skill set in leading change in HR group)*	*SVP of HR will meet with other HR leadership in order to engage in initiative; ensure understanding of "why" as well as discuss expected challenges* *GOAL: increase comfort level of HR department*	*Change Manager will work with SVP of HR to prepare for initial face-to-face meeting with HR leadership*
EXAMPLE: Internal Communi-cations	*Fitting communications about initiative in the already developed annual communication plan (focused on a number of upcoming change initiatives)*	*Medium involvement* *Need development of communication strategy and implementation of that strategy*	*Need budget for temporary resources (potentially to manage another major communication within the organization)*	*SVP Marketing & Communications will hold meeting with Internal Communications group to understand challenges.*	*N/A – no involvement from change team members needed yet*

Figure 10.6   Example stakeholder management matrix

This particular example focuses on the time period prior to the launch of the strategic initiative. A matrix must be developed for five time periods:

1. Prior to the launch
2. At launch
3. Throughout the implementation
4. After implementation/roll out
5. After initial measurement of progress toward achieving the goal

The information in the stakeholder management matrix is updated as needed and feeds into the communication plan (which will be discussed in the next section). A stakeholder management matrix puts the focus on key stakeholder groups which enables:

- Considering and taking into account the impact on the stakeholder
- Ensuring that stakeholders are engaged to address their specific needs
- Learning where challenges may exist that were unknown
- Finding champions as well as resisters to the initiative

Conversations with employees at all levels feed into the stakeholder management matrix, as well as the leaders' previous knowledge of stakeholders. This template is also available for download from the Web Added Value™ Resource Center at the www.jrosspub.com/wav.

Sharing the Plan, Getting Input, and Communicating

Figure 10.7 provides a flow for sharing the initial strategic plan, getting input from throughout the organization, and finalizing the plan.

While input to the plan is desired in the early stages of developing the plan—usually along the lines of beginning to sell the key objectives of the plan—once an initial draft is developed, all or some of it should be shared around the organization in order to gather feedback. This serves a number of purposes, including:

- Engaging employees in the strategic plan
- Identifying champions and resisters to the strategic plan
- Getting value insight from employees
- Enabling employees to feel a part of a bigger picture; involved in the direction of the organization

Not every organization shares the strategic plan in its entirety. Strategic plans often include sensitive information that leadership may

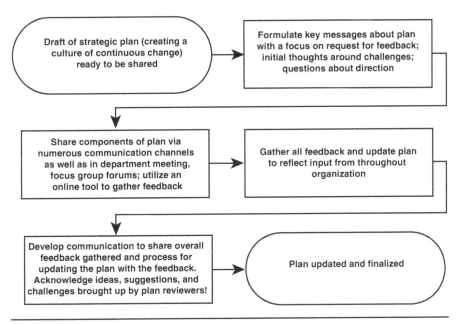

Figure 10.7   Communication flow: sharing the draft plan and getting feedback

not want broadly shared for fear of it getting into the hands of com-
petitors. However, it is essential to share components of the plan
that are focused on strategic goals since employees will need to work
toward achieving those goals.

A variety of channels should be used for sharing the plan. Prior to posting a plan on an internal site or sharing it via e-mail, first an all-staff or all-hands meeting should be held to share the key points of the plan. This enables executives to talk through the plan prior to letting employees read through it on their own. Taking this step before sharing the draft plan via other channels provides employees perspective and background information about the plan and its development.

Consider the use of a portal, such as Microsoft SharePoint™ to
share the plan and enable employees to provide feedback, as well as
collaborate and communicate with each other and leadership about
the plan.

Analyze and evaluate feedback gathered from employees' review of the strategic plan to determine what should, and can, be incorporated into the final plan. As a best practice, allocate feedback into three buckets:

1. Add to plan (*aligns to vision, mission, and core values; in alignment with organizational goals*)
2. Hold for further discussions (*not an exact alignment, but a suggestion worth pursuing further*)
3. Do not add to plan (*does not align to vision, mission, and core values; not in alignment with organizational goals*)

When possible, reach out to individuals who provided feedback to thank them for that feedback. When a decision has been made as to whether or not to incorporate the feedback, be sure to share what decisions were made and why. For those suggestions being put on hold for further discussions (bucket #2 in the list above), consider having the employee involved in those discussions. Should an initiative be launched based on further exploration of the suggestion, be sure to get the individual who proposed the idea involved in the project.

> *In one organization with which I work on annual planning sessions, the CEO writes a personal "thank you" note to each individual who provides detailed feedback on the strategic plan.*

Figure 10.8 provides a more detailed flow for communicating on the final strategic plan.

Figure 10.8 provides a processed approach to ensuring that the final plan is communicated in a way that reaches a broad stakeholder group, and communications remain consistent as implementation of the plan gets underway.

Once an organization has successfully created and nurtured a culture of continuous change, leaders find that employees regularly change how the work is done, often as a part of their everyday job.

> *After investing nearly three years in working toward a culture where change was the norm, one organization found that, over time, change was no longer a big deal in the organization. What this meant was that the ways of getting work done was regularly being evaluated and refined by employees. As new hires came into the organization, employees would reach out to them and ask them how the processes worked in their previous organization. They actively sought out best practices from others and incorporated those best practices into their current processes and procedures.*

Larger organizational change initiatives that are launched—such as a restructuring of the organization after a merger and acquisition, or centralizing project management offices into one enterprise project management

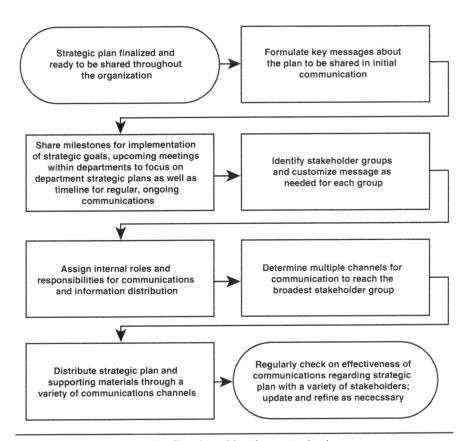

Figure 10.8   Communication flow: launching the strategic plan

office—should still be incorporated into the strategic plan of the organization and launched as separate, major change projects. However, those smaller changes that allow for continuous changes as to how work gets done will occur on an ongoing basis.

Remember, deciding to move to an organization that supports, encourages, and evaluates based on change as the norm does not happen overnight. Most organizations need to consider many other initiatives that must be undertaken in order to ensure the organization and its employees are set up for success. This might include:

- Restructuring the organization to allow for increased collaboration
- Ensuring guidelines around problem solving and making decisions
- Restructuring performance management systems to acknowledge continuous improvement involvement and compensate accordingly
- Increasing confidence and comfort in risk taking among employees

- Ensuring resources, budgets, and time allocated to work on continuous improvement
- Identifying KPI to measure the success of continuous improvement initiatives

To that end, a strategic plan initiative to move toward a culture of change as the norm will certainly be more than one year in duration!

> *A global application development firm launched an initiative to adopt a culture of regular and consistent change. The initial project of planning, through to implementation, took three years before leadership saw some positive results. Some key challenges that were addressed in the implementation of this strategic objective included:*
>
> - *Launching technology that would enable knowledge sharing throughout the organization*
> - *Adapting the performance management system to incorporate work on continuous improvement initiatives*
> - *Working with managers to ensure that all employees had up to six weeks per year "off line" to work on continuous improvement initiatives*
> - *Restructuring compensation and benefit packages to acknowledge and reward teamwork on continuous improvement initiatives*
>
> *At the time of the implementation of the continuous and regular change initiative, there was no project management office (PMO) in place. However, around the fourth year of the initiative being implemented, a group of cross-functional employees approached leadership with a proposal to create a centralized PMO in order to put more structure around, not just the continuous improvement initiative being undertaken in the organization, but also around projects in general. While there was much sharing of best practices and lessons learned between continuous improvement teams, it was felt that employees could be even more successful with a dedicated group to help coordinate and manage all initiatives. A new strategic initiative was launched with the goal of creating and launching a centralized PMO.*

This book has free material available for download from the
Web Added Value™ resource center at *www.jrosspub.com*

SUMMARY

"The rate of change is not going to slow down anytime
soon. If anything, competition in most industries will probably speed up even more in the next few decades."
John P. Kotter, *Leading Change*

Organizational change is not easy. It takes commitment—not just from
leadership but from *every* employee in the organization. It requires engaging people to help them to understand *why* the change is happening
and *why* it matters for the organization and for them individually. Change
requires changing behaviors—including how the individuals get the work
done and how they collaborate within the organization.

Change *must be aligned* to the current culture of the organization. If the
culture does not support change, then the culture must be changed for
any other change to be successful. If this is necessary, it is a long (but not
impossible) effort and requires significant investment in time, money, and
communications.

As has been mentioned throughout the book, engaging employees in
change is *the only way* to assure success of a change initiative. This requires
communication specific to:

- Planning for the change initiative
- Launching the change initiative
- Implementing the change initiative
- Evaluating the success of the change initiative

Consider this story of an organization who communicated up front in a variety of ways to engage employees prior to launching major change initiatives:

> *A global marketing and public relations firm was launching a large
> change initiative that would impact nearly every functional area in the
> organization. The initiative was coming on the heels of a merger and
> was being launched by the new Chief Executive Officer (CEO). The
> change included restructuring the organization to merge a number of*

305

departments; and then evaluating processes in the merged departments to find efficiencies. It would also entail redefining roles and responsibilities to enable for consistencies between the two merged organizations. Long before the change initiative was launched, the previous CEOs of both organizations reached out to their employees in a jointly developed communication. The purpose of the communication was to notify employees of the upcoming merger. It included information about the "why" and "when" of the merger as well as letting individuals know that it would take a couple of years to complete it (this was meant to reassure employees that no one was losing their job in the near future). The communication also listed change initiatives that would be launched—merger of departments, refined and updated processes, as well as redefined roles and responsibilities. This communication was followed up with a schedule of small group focus meetings, department meetings, as well as all-staff meetings to be held virtually so that all employees (from both organizations) could participate. This first communication noted that more information would be forthcoming and provided by division and department heads and that a portal would be launched that would be shared by both organizations—enabling collaboration between employees. Follow-up communications included information about:

- *Accessing the portal and training for use of the portal*
- *Upcoming meetings to discuss the merger*
- *Announcement of a new CEO*

Change initiatives, when launched, were done with the majority of employees from both organizations onboard and excited about the future. The CEO and other senior leadership attributed this to engaging employees very early on and addressing what they knew would be concerns.

As discussed throughout this book, organizational change *must be planned* if it is going to be successful—no matter how simple or complex the perception of the change. The more effectively leadership communicates about the change—via a number of communication channels—the more likely they will see a successful end result. Organizational change initiatives that are not successful will have an impact in a number of areas including:

- Reduced productivity due to reduced employee engagement, which causes:
  - ◊ Reduced revenue
  - ◊ Increased costs
  - ◊ Decreased customer satisfaction

- Negative gossip
- A lack of trust between employees and leadership
- The potential of the organization to lose top talent

Taking a strategic project management approach to organizational change provides increased structure behind the change initiative. This structure comes in the form of developing a scope of the change initiative, ensuring a sponsor to support the initiative, a project plan to manage against, assigning team roles and responsibilities for completing tasks, and developing a communication plan, as well as capturing lessons learned so that each successive change initiative is more successful than the last.

Throughout this book a number of best practices were provided on how to be successful in implementing organizational change that will be positive, help move the organization forward, and engage employees in the process. Remember, however, it is not a simple matter of just deciding to launch a change initiative and then doing so. Proper planning in the form of developing a strategy to implement organizational change is essential. Part of this strategy entails creating and sharing a vision for change that is beyond the benefits of change for the organization and includes the value of change *for each individual employee*.

In summary, each organizational change initiative must:

- Have a clear and concise vision for change supported by leadership that is shared across the organization
- Be aligned to the strategic plan of the organization
- Include a plan to gain consensus for the change from throughout the organization (create champions of change)
- Include a plan to train individuals in the change so that they have the skills and knowledge they need to be successful in the new environment
- Use a variety of communication channels to ensure that the broadest group of stakeholders is able to participate in the change initiative— whether through working on the change initiative, being part of a Stakeholder Support Committee, or providing feedback and input
- Include a variety of check-in points during and after implementation to ensure that the change is going as originally planned

To get started, visit the Web Added Value™ Download Resource Center at www.jrosspub.com/wav and download a number of tools, templates, and surveys to help frame the organization's next change initiative. These tools and templates will enable leaders to get their organizational change initiative started in the right way to increase chances for success.

Thank you for reading!

This book has free material available for download from the
Web Added Value™ resource center at *www.jrosspub.com*

INDEX